# PROGRESS IN BIOORGANIC CHEMISTRY

# PROGRESS IN BIOORGANIC CHEMISTRY

VOLUME FOUR

Edited by

## E. T. KAISER

*Departments of Chemistry and Biochemistry*
*University of Chicago*

## F. J. KEZDY

*Department of Biochemistry*
*University of Chicago*

A WILEY-INTERSCIENCE PUBLICATION

JOHN WILEY & SONS

NEW YORK · LONDON · SYDNEY · TORONTO

Copyright © 1976, by John Wiley & Sons, Inc.

All rights reserved. Published simultaneously in Canada.

No part of this book may be reproduced by any means, nor transmitted, nor translated into a machine language without the written permission of the publisher.

Library of Congress Cataloging in Publication Data:

Main entry under title:

Progress in bioorganic chemistry.

(Progress in bioorganic chemistry series)
Includes bibliographies.
1. Biological chemistry—Collected works. 2. Chemistry, Physical organic—Collected works. I. Kaiser, Emil Thomas, 1938-     ed. II. Kézdy, F. J., 1929-     ed.

QD415.A1P76          547'.1'3          75-142715
ISBN 0-471-45488-7(V.4)

Printed in the United States of America

10 9 8 7 6 5 4 3 2 1

# CONTRIBUTORS

THOMAS C. BRUICE, *Department of Chemistry, University of California, Santa Barbara California*

K. T. DOUGLAS, *Department of Chemistry, Duquesne University, Pittsburgh, Pennsylvania*

E. T. KAISER, *Department of Chemistry, University of Chicago, Chicago, Illinois*

F. J. KÉZDY, *Department of Biochemistry, University of Chicago, Chicago, Illinois*

P. A. LOACH, *Department of Biochemistry and Molecular Biology, Northwestern University, Evanston, Illinois*

# FROM THE
# PREFACE TO THE SERIES

Bioorganic chemistry is a new discipline emerging from the interaction of biochemistry and physical organic chemistry. Its origins can be traced to the enzymologists whose curiosity was not satified with the purification and the superficial characterization of an enzyme, to the physical organic chemists who had the conviction that the elementary steps of biological reactions are identical to those observed in organic chemistry, and to the physical and organic chemists who wished to understand and to imitate *in vitro* the unequaled catalytic power and specificity exhibited by living organisms. As with all interdisciplinary sciences, bioorganic chemistry uses many of the methods and techniques of the disciplines from which it is derived; many of its protagonists qualify themselves as physical organic chemists, enzymologists, biochemists, or kineticists. It is, however, a new science of its own by the criterion of having developed its own goals, concepts, and methods.

The principal goal of bioorganic chemistry can be defined as the understanding of biological reactions at the level of organic reaction mechanisms, that is, the identification of the basic parameters which govern these reactions, the formulation of quantitative theories describing them, and the elucidation of the relationships between the reactivity and the structures of the molecules participating in the process. This definition is narrower than one which some scientists would give. They might prefer to include areas such as medicinal chemistry, for example, as part of the field of bioorganic chemistry. Accordingly, the goals which they would cite would differ from those which we have considered. We do not seek here to argue or to defend our concept of what constitutes bioorganic chemistry, but within the framework of our definition we believe that there is a real distinction between much work in present day medicinal chemistry and that in the bioorganic field. In our conception of bioorganic chemistry the emphasis is on mechanism.

The theoretical formulation of the understanding of biochemical and, therefore, enzyme-catalyzed reactions has required the elaboration of new concepts, such as multifunctional catalysis, stereospecificity by three-point attachment, and control of reactivity by conformational changes. Many of the new concepts will not survive; they will be redefined, discarded, or reevaluated as fortunately always happens in science. But the trend is clearly apparent—these new concepts are providing us with efficient tools of great power which can be used to describe and discuss enzymatic reactions.

As to the methods of bioorganic chemistry, they are conceptually the same as those for the study of any chemical reaction; they include analytical and physical techniques. However, the complexity of the reacting molecules has resulted in methods which are new and unique in their ability to probe the chemistry of a functional group surrounded by a multitude of very similar groups or a chemical event accompanied by a host of satellite reactions. The discovery of numerous methods involving active-site directed reagents, "reporter molecules," and chromophoric substrates illustrates the usefulness and the elegance of the new science.

The future of bioorganic chemistry appears very promising, and the fields to cover in the future are immense and unexplored. The earliest work has concentrated on the understanding of general acid—general base catalyzed reactions, hydrolytic reactions, and the role of proteins in enzymatic catalysis. The mechanism of enzymatic catalysis by most coenzymes is very far from being well described, and the very prominent role of metal ions in catalysis is only beginning to emerge. Other important problems, such as surface catalysis at biological membranes, transport mechanisms, the process by which ribosome-catalyzed reactions occur, and the reactivity of RNA and DNA molecules, are at an early stage of development or are completely unexplored.

As a result, a rapid growth of bioorganic chemistry is desirable and is currently underway, as evidenced by the large number of papers published on the subject. An unfortunate result of this rapid expansion is the scarcity of comprehensive treatments of bioorganic chemistry. The rapid progress in this field makes it likely that large portions of any comprehensive textbook will become obsolete soon, although the student of bioorganic chemistry may still learn some of the basic concepts of the subject from them. Because of many factors, it would be possible to revise textbooks only at infrequent intervals. For this reason the format of presenting comprehensive treatments of limited subjects seemed more appropriate to us. It would provide the investigators, interested readers, and students with a thorough and critical evaluation of those aspects of bioorganic chemistry where definite and substantial progress has been achieved.

It is the hope of the Editors of this series to be able to respond to the need for up-to-date comprehensive treatments of important topics in bioorganic chemistry. In attempting to do so we would like to provide treatments of bioorganic subjects which will be general enough to retain the attention of most workers in the field and which will be at a level beyond that of a usual review article or literature survey. Since many aspects of bioorganic chemistry are still in the process of evolution, we also would like to provide a forum where the authors can express challenging new ideas and present stimulating and, frequently, controversial discussions. For this reason we hope to give the authors somewhat more latitude than is customary in this kind of publication, while still retaining the requirement of scientific sobriety.

E. T. KAISER
F. J. KÉZDY

*Chicago, Illinois*
*January 1971*

# PREFACE TO VOLUME FOUR

The proceding volumes of *Progress in Bioorganic Chemistry* have dealt heavily with the mechanism of action and the structure of proteins, including such species as metalloenzymes. Because of their catalytic role in most biochemical reactions, enzymes are still the focus in bioorganic chemistry. The present volume diverges somewhat from this emphasis. The chapters on photosynthesis explores the importance of supramolecular structures and the sophisticated experimental approaches necessitated by the complexity of such structures. The three other chapters, however, deal with intimate aspects of reaction mechanism at the level of small organic molecules of biological interest. From the studies of the small molecules described it is hoped that much that is relevant to the understanding of more complex biological systems will emerge.

T. E. KAISER
F. J. KÉZDY

*Chicago, Illinois*
*February 1976*

# CONTENTS

# MODELS AND FLAVIN CATALYSIS

THOMAS C. BRUICE

*Department of Chemistry*
*University of California*
*Santa Barbara, California*

1

# 1  INTRODUCTION

Flavins share in common the isoalloxazine ring structure which may exist in oxidized, radical, and reduced forms as shown in Scheme 1 [1-8, 8a-c] for a typical 7,8-dimethyl 10-alkyl or ribitol isoalloxazine. In Scheme 1 the values of $\lambda_{max}$ and $\epsilon$ should not be taken at face value for all species, since literature sources differ in conditions and exact assignments and these constants are often sensitive to solvent species present. The flavin species encountered at other than the extreme ends of the pH range are the neutral oxidized flavin ($F_{ox}$), the neutral 1,5-dihydroflavin (FH$_2$) or its conjugate base formed by ionization of the N(1)-proton (FH$^-$), and the three radical species as abbreviated in Scheme 1. In most mechanistic studies the N(3)-proton is replaced by an alkyl substituent in order to obviate the added complexity to the pH dependence of rate brought about by N(3)-H ionization when the flavin is in the oxidized state.

Oxidized and reduced flavin comproportionate to provide flavin radical through charge transfer species as shown in eq. 1 [9,10] for flavin mononucleotide (pH 4.5, 11°C). Determination of the

$$\text{FMN} + \text{FMNH}_2 \xrightleftharpoons[>10^6 \text{ sec}^{-1}]{>10^8 \ M^{-1} \text{ sec}^{-1}} \text{FMN} \cdot \text{FMNH}_2$$

$$\text{FMN} \cdot \text{FMNH}_2 \xrightleftharpoons[1 \times 10^7 M^{-1} \text{ sec}^{-1}]{3.5 \times 10^3 \text{ sec}^{-1}} 2 \text{ FMN} \cdot$$

$$(1)$$

individual rate constants of eq. 1 is complicated because of an energetically favorably dimerization of the FMN· species ($K_e \cong 5 \times 10^{-7}$

$M$) [10]. The mole fraction of radical species present in aqueous solution of oxidized and reduced flavin increases at high and particularly at low pH (eq. 2) [3]. Since variously modified isoalloxazines

$$K_T = \frac{[FR\cdot]^2_{total}}{[FH_2]_{total}[F_{ox}]_{total}}$$

(2)

$$\overline{K} = \frac{[FRH]^2}{[FH_2][F_{ox}]}$$

$\lambda_{max}$ 390 ($\epsilon = >2 \times 10^4$)　　$\lambda_{max}$ 446 ($\epsilon = 1.25 \times 10^4$)　　$\lambda_{max}$ 444
　　　264　　　　　　　　　　　　370 ($\epsilon = 9.8 \times 10^3$)　　　　　　350
　　　　　　　　　　　　　　　　270 ($\epsilon = 3.96 \times 10^4$)　　　　270

$F_{ox}H^+$ 　　　　　　　　　　$F_{ox}$ 　　　　　　　　$F_{ox}^-$

$+e$ ‖ $-e$ 　$E_0 = -0.24$ V
$+H^+$ ‖ $-H^+$

$\dot{F}RH_2$ red 　　　　　　　$\dot{F}RH$ blue 　　　　$\dot{F}R^{\ominus}$ red
$\lambda_{max}$ 350 ($\epsilon = 13,000$)　pH 5.1 　　　　pH 11.4
　　　490 ($\epsilon = 12,300$)　$\lambda_{max}$ 570 (5130)　$\lambda_{max}$ 370 ($\epsilon = 16,000$)
at $H_0 = -1.1$ 　　　　　　　　　　　　　　　　400
　　　　　　　　　　　　　　　　　　　　　　480 ($\epsilon = 4000$)

$+e$ ‖ $-e$ 　$E'_0 = -0.17$
$+H^+$ ‖ $-H^+$

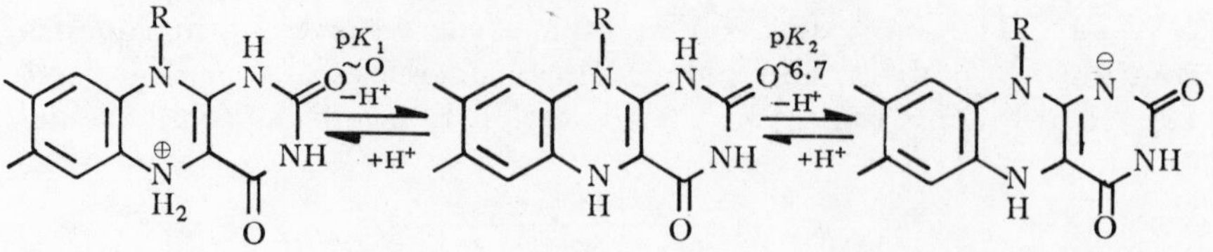

$FH_3{}^{\oplus}$
6N HCl
$\lambda_{max}$ 250
316

$FH_2$
pH2.0 $SO_4{}^{2-}$buffer
$\lambda_{max}$ 250
~280 shoulder
~400

$FH^{\ominus}$
pH 9.0 Borate buffer
$\lambda_{max}$ 256
288 shoulder
~ 350

Scheme 1

may have potentials and $pK_a$ values associated with their oxidized, reduced, and radical species quite divergent from those for normal flavins, the pH dependence for the equilibrium concentration of the radical species is expected to vary considerably. By defining [10a] the equilibrium constants of eq. 2 and assuming the mole fraction of the $FH_2 \cdot F_{ox}$ complex to be negligible, we obtain eq. 3. The values of $K_{F_1}$, $K_{F_2}$, $K_1$, and $K_2$ (Scheme 1) may be obtained by spectro-

$$K_T = \bar{K} K_1 K_{F_1} \text{x}$$

$$\left\{ \frac{[K_{R_1} K_{R_2} + K_{R_1} H^+ + [H^+]^2)^2}{K_{R_1}^2 (K_1 K_2 + K_1 H^+ + [H^+]^2)(K_{F_1} K_{F_2} + K_{F_1} H^+ + [H^+]^2)} \right\} \quad (3)$$

photometric titration of the oxidized and reduced isoalloxazine and $K_T$ may be computer solved by iteration for $\bar{K}$, $K_{R_1}$, and $K_{R_2}$. When the N(3)-position is alkylated, as is the case in most physical organic studies, eq. 3 simplifies to eq. 4.

$$K_T = \frac{K_{pH} K_1 K_{F_1} (K_{R_1} K_{R_2} + K_{R_1} H^+ + H^{+2})^2}{K_{R_1}^2 H^+ (K_1 K_2 + K_1 H^+ + H^{+2})(K_{F_1} + H^+)} \quad (4)$$

Flavins reduce at the dropping mercury electrode in 2 one-electron steps that are separated in aqueous acid [11] but that merge into a nonideal wave at alkaline pH values. The half-height of the wave $(E_{\frac{1}{2}})$ may be employed [12] (much as a Hammett sigma value [13]) as an index of electron availability since $\Delta F^0$ for $F_{ox} \rightarrow FH_2$ + $FH^-$ is ideally related to $E_{\frac{1}{2}}$. As a planar polycyclic molecule, $F_{ox}$

is prone to form face-to-face complexes in aqueous solution (for a compilation of pertinent references see reference [14]. Complex formation, both productive and nonproductive, must be considered as a part of certain $F_{ox}$ catalyzed reactions. Massey and Ghisla have recently [15] summarized the evidence indicating that complex formation (with resultant charge transfer interaction) is an important feature of the mechanism of many flavoenzymes. A free energy relationship that is useful to test the importance of complex formation involves a plot of log $k_{rate}$ versus log $K_e$, where $K_e$ represents the equilibrium constant for 1:1 complex formation of each of a series of isoalloxazines with tryptophan or $\beta$-resorcylic acid [13]. The relationships of log $k_{rate}$ to $E_{1/2}$ and log $K_e$ have proven useful in studies of the reaction of $SO_3^{2-}$, thiols, carbanions, and dihydropyridines with oxidized isoalloxazines. The usefulness of linear free energy plots based on $E_{1/2}$ and log $K_e$ for mechanistic deduction await (1) the determination of these values for a large number of N(3)-alkyl isoalloxazines and (2) application of such plots to a sufficient number of reactions to allow meaningful interpretations of slopes. The use of $E_{1/2}$ values rather than $E_0'$ is certainly allowed and has the advantage of the ease of determination of the former (for use of $E_0'$ see Müller and Massey [16]). Mechanistic studies are made operationally more simplistic if restricted to N(3)-alkyl isoalloxazines since ionization of the N(3)-H unduly complicates the interpretation of log $k_{rate}$ versus pH profiles, and so on. The substitution of alkyl groups for hydrogen on the N(3)-position removes the flavin anion species from consideration. This is probably not important in studies of flavin catalysis, though ionization of the N(3)-proton does change the susceptibility of the isoalloxazine ring to nucleophilic attack (see hydrolysis p. 18). In Table 1 are collected the values of $E_{1/2}$ and $K_e$ for a series of N(3)-alkyl isoalloxazines.

While $F_{ox}$ is a planar molecule, the $FH_2$ molecule resembles a butterfly in conformation with the atoms 1-5 and 10, and 5-9 and 10 constituting two planes (wings) intersecting in the N(5)-N(10) axis with a dihedral angle of $21°$ (I). In the case of N(5) substituted flavins the angle may vary widely (9-36°). A collection of pertinent

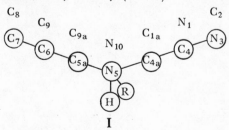

I

TABLE 1 POLAROGRAPHIC $E_{1/2}$ VALUES (30.0°C, pH 8.97) WITH RESPECT TO SATURATED CALOMEL ELECTRODE AND EQUILIBRIUM CONSTANTS FOR 1:1 COMPLEXING WITH TRYPTOPHAN OR β-RESCORCYCLIC ACID AT 22°C, pH 7.85

| X | R2 | Y1 | Y2 | Y3 | $E_{1/2}$ (V) | $K_e$ Tryptophan ($M^{-1}$) | β-Resorcyclic Acid ($M^{-1}$) | Reference |
|---|---|---|---|---|---|---|---|---|
| CH | CH3 | H | H | H | (−0.805)[a] | 102[b] | 83[b] | 18 |
| N | CH3 | H | H | H | −0.480 | 86 | 57 | 19 |
| N | CH3 | H | Cl | H | −0.445 | 118 | 84 | 19 |
| N | C6H5 | H | H | H | −0.450 | 47 | 40 | 19 |
| N | o-CH3−C6H4 | H | H | H | −0.445 | 32 | 34 | 19 |
| N | 2',6'-(CH3)2−C6H3 | H | H | H | −0.440 | 19 | 14.3 | 19 |
| N | 2',6'-(CH3)2−C6H3 | SO3H | H | SO3H | −0.345 | c | c | 18 |
| N | CH3 | H | CH3 | CH3 | −0.510 | 113 | 85 | 18 |
| N | CH3 | H | H | CN | −0.433 | | | 20 |
| N | CH3 | H | CN | H | −0.333 | | | 20 |

[a] The value for the 5-deazaflavin may not be related to $E_0'$ in the same manner as is the case of flavins. A $E_0'$ value of −315 mV has tentatively been suggested by G. Blankenhorn (17b).
[b] 25.3°C.
[c] Too low to be determined.

references may be found in [17], see also [17a]. The rate of inter-conversion of the two possible butterfly conformations has been measured for several reduced flavins [17] by nmr observation of the pyramidal inversion of the N(5)-center. The enthalpy of the inversion barrier was found to be ~10 kcal mole$^{-1}$ in acetone, increasing to ~13 kcal mole$^{-1}$ in aqueous solution. It has been suggested by Hemmerich et al. [17] that the $E_0'$ of flavoenzymes could be controlled by the extent of preferential binding of the apoprotein to the flat or the butterfly conformation of $F_{ox}$ and $FH_2$, respectively. Thus, enforced flattening of the reduced form should increase its reducing power and conversely, enforced bending should enhance the oxidizing power. Selectivity for $1e^-$ and $2e^-$ reactions could conceivably be assisted in the same manner. Recent crystallographic data at 1.9 Å with clostridial flavodoxin has been interpreted to support the concept [21] that the protein tertiary structure can control cofactor conformation, since in reduced, oxidized, and semiquinone states there are no major changes in protein structure. The suggestion has been made that in all states of oxidation, the flavin moiety of the flavodoxin is planar. Flavin radical has been shown to complex with tryptophan and tyrosine in aqueous solution [22]. The complex with tryptophan, but not those with tyrosine or methionine, has been found to have approximately one-tenth the rate constant in reaction with various oxidizing agents [23]. It is interesting to note that the flavodoxins function by shuttling between the semiquinone and reduced state and that tryptophan is the conserved amino residue in the coenzyme binding sites of several flavodoxins. The decrease in rate constants noted on complexation of flavin radical to tryptophan may be due to several factors, including steric hindrance to approach of oxidant, change of conformation of the radical moiety, and electronic interaction of complexing species.

The biocatalytic flavin entities are most usually flavin mononucle-otide (FMN) and flavin adenine dinucleotide (FAD). For an

$$(CHOH)_3-CH_2-O-\overset{\overset{\displaystyle O}{\|}}{\underset{\underset{\displaystyle OH}{|}}{P}}-OH$$

FMN

FAD

increasing number of enzymes these cofactors are covalently attached to the enzyme at the 8-α-carbon through a linkage to the N(3) of a histidine imidazolyl group or the sulfur of a cysteine residue (for a short review see reference 24). Attachment of the 8-α-methylene group to the N(3) of histidine is found in succinate dehydrogenase [25, 26] and D-6-hydroxynicotinamide oxidase [27] and is likely to occur in sarcosine dehydrogenase [28]. Interestingly, the N(3)-isomer is also the initial product of the reaction of 8-α-bromotetraacetyl riboflavin and benzoyl histidine. Acid catalysis converts the N(3)- to the N(1)-isomer in an as yet little studied reaction (eq. 5). For thiamine dehydrogenase [29] and β-cyclopia-

$$(5)$$

zonate oxidocyclase [30], a histidine imidazolyl group is attached to the 8-α-position, but the linkage is apparently other than a Im—CH$_2$ bond. Nmr, hydrolytic, and other data are suggested to support a carbinolamine structure probably involving the N(3)-nitrogen (II). For mitochondrial monoamine oxidase, cysteine is attached to the C(8α)-

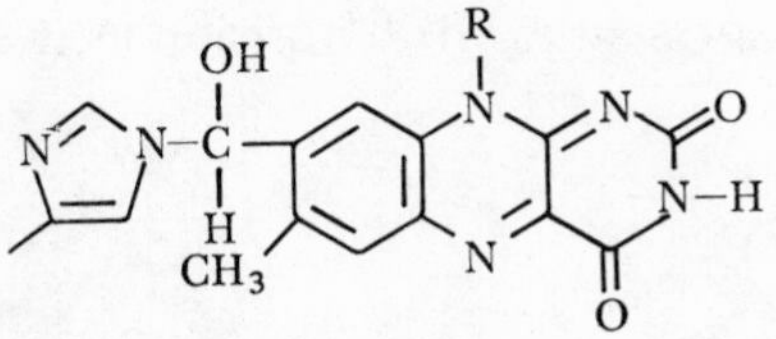

**II**

position by a thioether linkage, but for *chromatium* cytochrome $C_{552}$ bonding is different and is thought likely to represent a cysteinyl thiohemiacetal [31, 32[. Studies in Singer's laboratory [24] have been interpreted to indicate that 8-formyllumiflavin and 8-formylte-traacetylriboflavin form thiohemiacetals with cysteine and N-acetylcysteine in solution. Extra stability for the thiohemiacetal structure in chromatium cytochrome $C_{552}$ is suggested to be supplied by a neighboring aromatic ring of a tyrosine residue. Thus the peptic

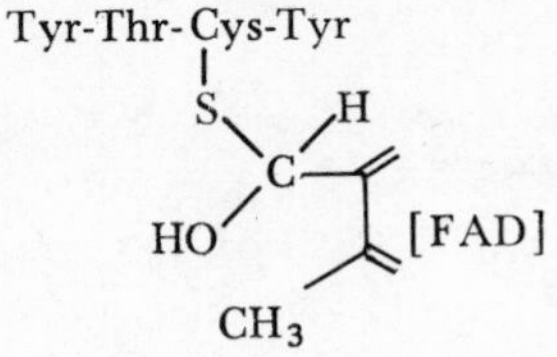

**III**

peptide (III) from *chromatium* cytochrome $C_{552}$ is reported to be

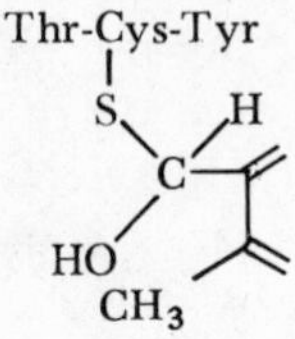

**IV**

far more stable than the corresponding trypsin-chymotrypsin peptide (**IV**) [24]. 8-Formylriboflavin exists as an internal cyclic acetal formed by addition of a ribitol hydroxyl group to the 8-formyl group [33].

In microorganisms a group other than methyl may be found at the 8-position of the flavin moiety of FMN or FAD (**V, VI**), or a hydroxyl group may be present at the 6-position (**VII**).

**V** [34-38]
$\lambda_{max}$ 472 nm ($\epsilon$ = 4.15 × 10⁴)
p$K_a$ 4.8

**VI** [39]
$\lambda_{max}$ 496 nm ($\epsilon$ = 3.9 × 10³)

R = ribityl-5′-ADP

**VII** [34-38] $\lambda_{max}$ = 650
($\epsilon$ = 2.3 × 10³) p$K_a$ ≈ 6.9

Although the photochemistry of the flavins has been extensively studied in the laboratories of Hemmerich, Tollin, and others and an extensive literature exists on the enzymology of flavoproteins, scant attention has been accorded to the type of physical organic studies that are required to elucidate the mechanisms of reactions (dark) of flavins. The progress in this area of research is summarized in this review and the author presents his views on flavin reaction mechanisms. Before proceeding, however, it is useful to consider some of the types of reactions catalyzed by flavoenzymes in order to perceive some of the types of mechanistic questions that must be asked. These enzymatic reactions (or their reverse) include dehydrogenations (eq. 6-11) and reactions with oxygen species (eqs. 12-14) as exemplified by the various oxidases. Equations 6 to 14 are not a

$$F_{ox} + 2\,RSH \longrightarrow FH_2 + RSSR \tag{6}$$

$$F_{ox} + \;\underset{\diagup}{\overset{\diagdown}{C}}\!\!\overset{H}{\underset{\diagdown}{\;}}\!\!-\!\!\overset{H}{\underset{\diagup}{\;}}\!\!\underset{\diagup}{\overset{\diagdown}{C}} \longrightarrow FH_2 + \underset{\diagup}{\overset{\diagdown}{C}}\!\!=\!\!\underset{\diagup}{\overset{\diagdown}{C}} \tag{7}$$

$$H^+ + F_{ox} + \underset{\underset{R}{N}}{\overset{\overset{H\quad H}{\diagup}}{\bigcirc}}CONH_2 \longrightarrow FH_2 + \underset{\underset{R}{\overset{\oplus}{N}}}{\bigcirc}CONH_2 \qquad (8)$$

$$F_{ox} + H{-}\overset{|}{\underset{|}{C}}{-}OH \rightarrow FH_2 + {>}C{=}O \qquad (9)$$

$$F_{ox} + H{-}\overset{|}{\underset{|}{C}}{-}NH_2 \rightarrow FH_2 + {>}C{=}NH \qquad (10)$$

$$FH_2 + 5,10\text{-methylene-}H_4\text{-folate} \rightarrow F_{ox} + 5\text{-methyl-}H_4\text{-folate} \qquad (11)$$

$$FH_2 + O_2 \rightarrow FRH\cdot + O_2^{\bar{\cdot}} + H^+ \qquad (12)$$

$$FH_2 + O_2^{\bar{\cdot}} \rightarrow FRH\cdot + HO_2^- \qquad (13)$$

$$FH_2 + O_2 \rightarrow F_{ox} + H_2O_2 \qquad (14)$$

comprehensive listing of types of flavoenzyme-catalyzed reactions (see reference 40) but are those for which some bioorganic literature exists.

The oxidase reactions include (1) oxidative decarboxylation of α-hydroxy acids by internal mixed-function flavomonooxygenases (which at the active site probably involves [41, 41a] reduction of $F_{ox}$ by α-hydroxy acid (eq. 9) to yield $FH_2$ and α-keto acid, reaction of the resulting $FH_2$ with $O_2$ to yield $H_2O_2$ (eq. 14), and reaction of $H_2O_2$ with α-keto acid (eq. 15) [42]) and (2) external flavomono-

$$\underset{\underset{R}{|}}{\overset{O\diagdown\diagup O^{\ominus}}{\underset{O{=}C}{C}}} + HO_2^{\ominus} \overset{H^+}{\longrightarrow} \underset{\underset{R}{|}}{\overset{O\diagdown\diagup O^{\epsilon}}{\underset{{}^{\ominus}O{-}\overset{|}{\underset{|}{C}}{-}O{-}\overset{\oplus}{O}H_2}{C}}} \overset{-CO_2}{\underset{-H_2O}{\longrightarrow}} \underset{\underset{R}{|}}{\overset{{}^{\ominus}O\diagdown\diagup O}{C}} \qquad (15)$$

oxygenases that hydroxylate aromatic (and imidazolyl) ring systems ortho or para to a hydroxyl (or ${\geq}NH$) substituent group (eq. 16). The external flavomonooxygenases cannot be shown to provide

$$\underset{X}{\bigcirc}{-}OH + FH_2 + O_2 \overset{-X}{\longrightarrow} \underset{OH}{\bigcirc}{-}OH + F_{ox} + H_2O \qquad (16)$$

arene oxides as intermediates since their hydroxylation products do not exhibit evidence of the NIH-shift (see eq. 17). This does not prove that arene oxides are not formed, however, since an α-hydroxy arene oxide would be expected to open spontaneously (eq. 17). Mager and Berends [43] provide experimental evidence (substituent

$$\tag{17}$$

patterns, etc.) that in model systems the hydroxylating species (acidic media) formed on reaction of $O_2$ and $FH_2$ is HO· (see also reference 44). In model reactions the hydroxylating species may arise from the Haber-Weiss reaction (eq. 18) [45]. Hydroxylation of

$$O_2^{\cdot -} + H_2O_2 \rightleftharpoons HO· + HO^- + O_2 \tag{18}$$

aromatic compounds by HO· is well known [46-47] and is not characterized by giving NIH-shifted products. For an enzymatic hydroxylation the mechanism of eq. 18 suffers insofar as $O_2^{\cdot -}$ once formed, would be required to search through the solution for an $H_2O_2$ molecule generated at another site. The suggestion of HO· as the hydroxylating species must be weighed against the finding by Strickland and Massey [48] that, at neutral pH, generation of $O_2^{\cdot -}$ continuously by irradiating a solution of lumiflavin-3-acetic acid in the presence of $O_2$ hydroxylates $p$-hydroxybenzoate in the presence of catalase [48]. Presumably in such a system $H_2O_2$ would be scavenged and production of HO· (eq. 18) prevented. Addition of superoxide dismutase to the reaction mixture all but prevents the hydroxylation reaction. Therefore either catalase can provide some species with $O_2^{\cdot -}$ which can hydroxylate $p$-hydroxybenzoate (as does cytochrome P-450 [49]), or an $O_2^{\cdot -}$ flavin complex must be the hydroxylating agent, because infusion of electrolytically produced $O_2^{\cdot -}$ into solutions of $p$-hydroxybenzoate does not result in the production of dihydroxybenzoates [48]. Hamilton [50] is of the opinion that an "oxene" reagent formed on reaction of $O_2$ with $FH_2$ is the hydroxylating agent (eq. 19). This remains an interesting specula-

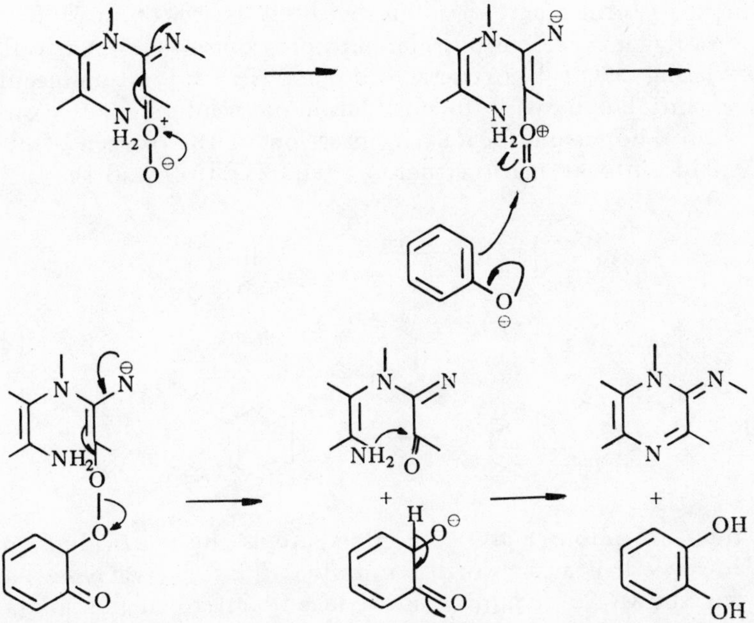

(19)

tion that suffers from the anticipation of a competing and irreversible intramolecular oxidation of the o-phenylenediamine moiety of the "oxene" intermediate. The mechanism of eq. 20 has been proposed by Orf and Dolphin [51]. Reduced flavin has long been

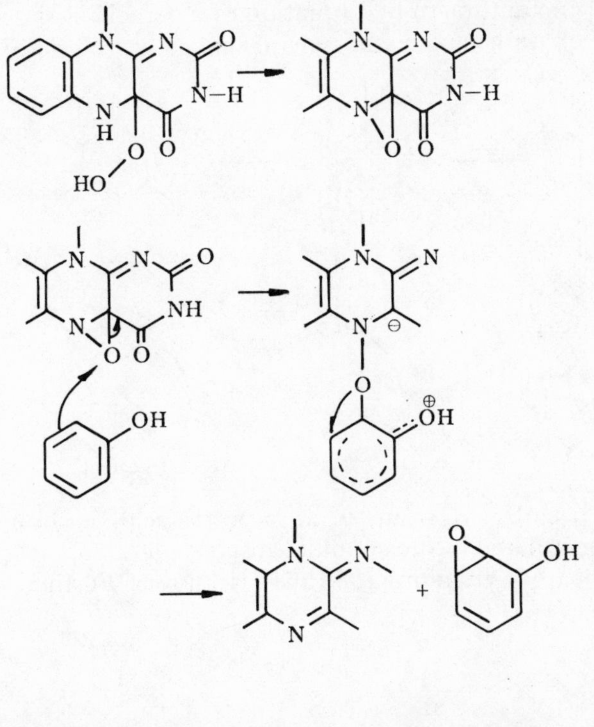

(20)

considered to form a peroxy adduct with $O_2$ (see Section 5). In eq. 20, the 4a-flavoperoxide is postulated to provide the 4a,5-oxaziridine hydroxylating agent by general-acid catalysis and intramolecular displacement. The basis of the postulation of an intermediate oxaziridine is the known photocatalytic insertion of the oxygen of pyridine-$N$-oxide into aromatic structures (eq. 21) [52] and the likeli-

$$\text{(21)}$$

hood that the photocatalytic intermediate is the oxaziridine [53, 54]. However it is again worth noting that the flavoxygenases have not been shown to require arene oxides as intermediates in their catalysis of hydroxylation reactions. Also, there is no evidence for a nonphotocatalyzed oxygen insertion reaction involving an oxiziridine. This fact would seem to remove the premise upon which eq. 20 is based. Obviously there is much to be done in this area of investigation. A simplistic mechanism for external flavomonooxygenase is suggested by the author to be formation of a peracid as shown in eq. 22. In this equation enzymatic general-acid-catalyzed peroxy anion

$$\text{(22)}$$

attack upon a carboxyl group of an aspartic acid or glutamic acid residue is postulated. Nucleophilic displacement upon carboxyl groups (even carboxyl anions [55, 56]) is known. Further, peracids are known hydroxylating agents and can insert oxygen to provide arene oxides [57-59]. This may also be a reasonable mechanism for the pteridine-requiring phenylalanine hydroxylase [60]. Phenylalanine hydroxylase is known to provide NIH-shifted products [60]. (For a kinetic study of the reaction of $O_2$ with tetrahydropteridines see reference 61).

Flavoproteins also perform important roles as one-electron transfer vehicles to nitrogenase, the respiratory chain cytochromes, and cytochrome P-450, P-445, and so on. The cytochrome P-450 systems are responsible for oxene insertions into biotic and exobiotic aromatic nuclei. The intermediacy of arene oxides was established by showing (eqs. 23 and 24) that the enzymatic hydroxylation reaction

$$O_2 + \quad \xrightarrow[\text{P-450}]{\text{cytochrome}} \quad \tag{23}$$

X = R, Cl, H, etc.

$$\xrightarrow{\text{NIH shift}} \tag{24}$$

and the rearrangement of arene oxides have in common the NIH-shift mechanism [62-66]. The mechanisms for the NIH shift have been established to involve specific acid catalysis and spontaneous carbonium ion formation followed by a group migration (eq. 25) [67-70].

$$\tag{25}$$

In the case of the cytochrome P-450 type enzymes, the flavoprotein only serves as an electron carrier and plays no direct role in transfer of an oxygen atom to substrate.

Bacterial luciferases are flavoproteins which yield light in their $O_2$ oxidation of long chain aldehydes (eq. 25A; see pg 79). The mechanism of eq. 25B has been suggested by Hastings and collaborators

$$FH_2 + O_2 \rightarrow FH_2O_2$$

$$FH_2O_2 + RCHO \xrightarrow{\;h\nu\;} F_{ox} + H_2O + RCO_2H \qquad (25A)$$

$$(25B)$$

[70A]. In 25B the collapse of the peroxy hemiacetal may be recognized as a Baeyer-Villiger type reaction. An alternate scheme that has occurred to the author involves imine formation with the 4a-peroxy adduct of $F_{ox}$ followed by cleavage of the resultant dioxetane (eq. 25C). The mechanisms of 25B and 25C rather resemble the proposed peroxy cleavage [DeLuca, 70B] and dioxetane [White,

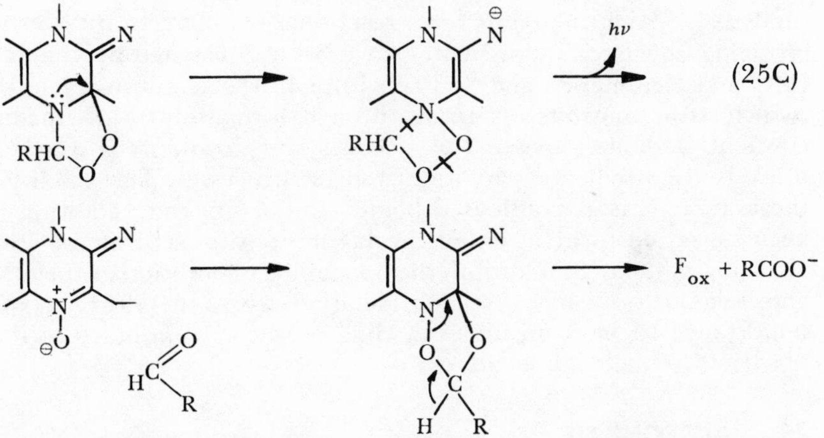

(25C)

$$F_{ox} + RCOO^-$$

70C] mechanism, respectively, for firefly luciferase. (Recently acquired information on the bacterial luciferase problem has been included on pg 79 to 80.)

In the two-electron oxidation or reduction of one organic substance by another, the electrons may be passed from reductant to oxidant in 2 one-electron steps (radical reaction) or an electron pair transfer may be involved. The latter process may occur by a direct transfer of hydride ion or in a two-step process involving in the first step nucleophilic addition of the reductant to the oxidant. The second step may then involve an E2 elimination (eq. 26) or a second

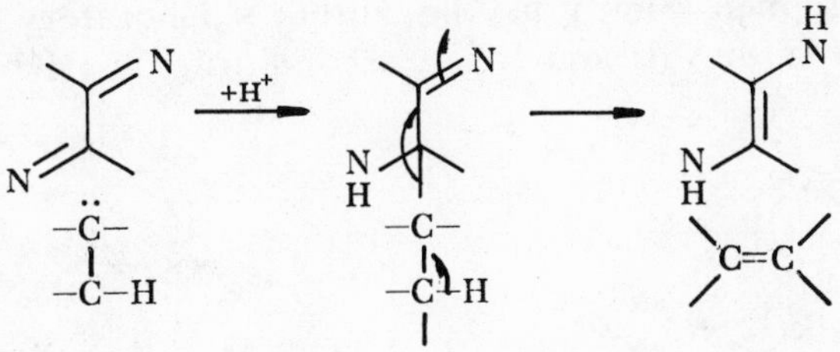

(26)

nucleophilic attack of reductant (eq. 27) [50, 71, 72]. For flavin-

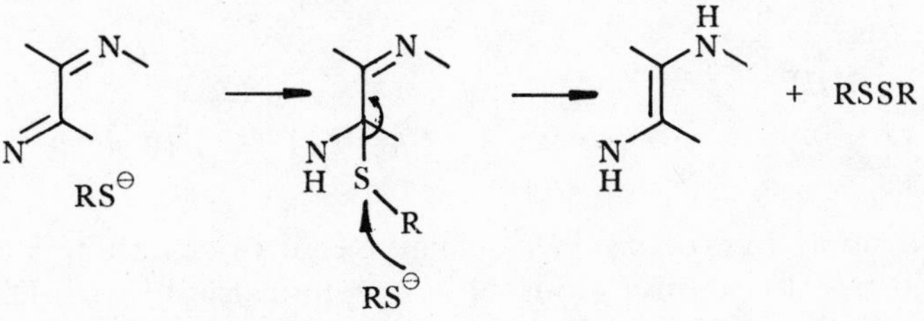

+ RSSR

(27)

catalyzed oxidation-reduction reactions, two-electron transfers involving covalent intermediates have been postulated by Hamilton [50] and Hemmerich and Shumen Jorns [73]. Hamilton argues that two-electron transfers occurring through formation of intermediate covalent adducts obviate the necessity of invoking energetically unlikely free radical or hydride transfer processes. The validity of these arguments is considered below. Initial substrate addition has been suggested to be at either the 4a- or 5-positions of the isoalloxa-zine ring system. In addition, the possibility of addition at the 6-, 8-, and 10a-positions requires consideration. It is therefore of some importance to have an understanding of the susceptibility of these positions to nucleophilic attack.

## 2 REACTIONS OF $F_{ox}$

### 2.1 Electrophilic Positions of the Isoalloxazine Ring System as Shown from Hydrolytic Studies

Hydroxide ion serves as a representative nucleophile, and informa-tion concerning the positions of its addition to the isoalloxazine ring system may be obtained from studies of the alkaline hydrolysis of substituted isoalloxazines. By this method one cannot detect even facile additions of hydroxide ion if they do not lead to ring opening, a very real possibility. An examination of this problem has been made by Stephen Smith in the author's laboratory [74]. The hydrolysis of both VIII and IX are second order in [HO⁻] indicative

|  |  |
|---|---|
| **VIII** | **IX** |
| $k[HO^-]^2 = 2.24 \times 10^2 \ sec^{-1}$ | $k[HO^-]^2 = 3.16 \times 10^2 \ sec^{-1}$ |

of amide bond hydrolysis [75]. That initial nucleophilic displace-ment is at the 4-carbonyl group is best demonstrated with IX, since the immediate product is the quinoxolone ureido imine X, which upon acidification reforms **IX** and upon admittance of $O_2$ provides the ring-contracted product **XI** (Scheme 2). In contrast to **VIII** and

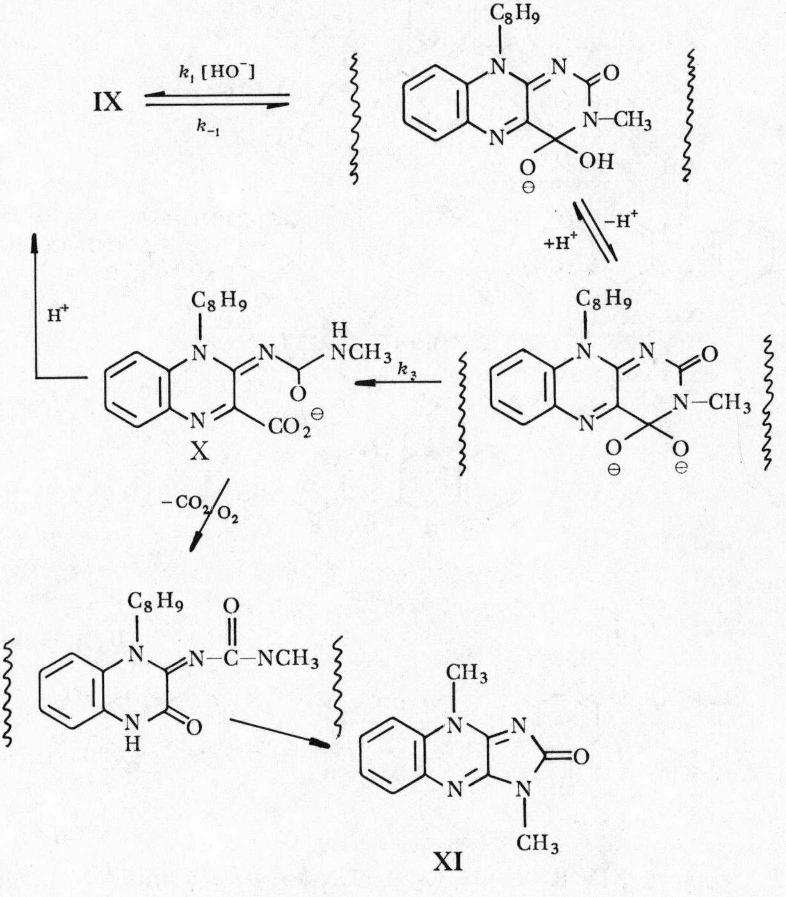

Scheme 2

and **IX, XII,** which is less sterically hindered at the 10a-position, undergoes hydrolytic scission at both the 4- and 10a- positions (Scheme 3). Removal of the 3-methyl group of **VIII** or **XII** provides acids of p$K_a$ ~9.5 (30°C). When the N(3)-proton has been ionized, the 10-phenyl compound hydrolyzes slowly at the 4-position in a reaction that is first order in hydroxide ion [74] while the 10-methyl compound hydrolyzes solely at the 10a-position [76-77]. Hydrolysis of the electron deficient isoalloxazine **XIV** and **XV** (eq. 28) provides the corresponding spirohydantoins by way of a reaction that is first

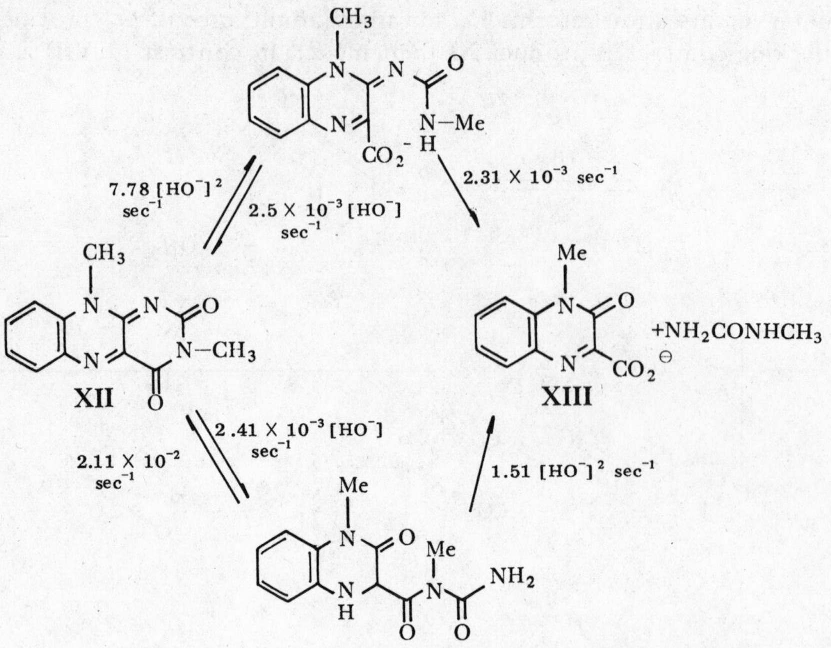


Scheme 3

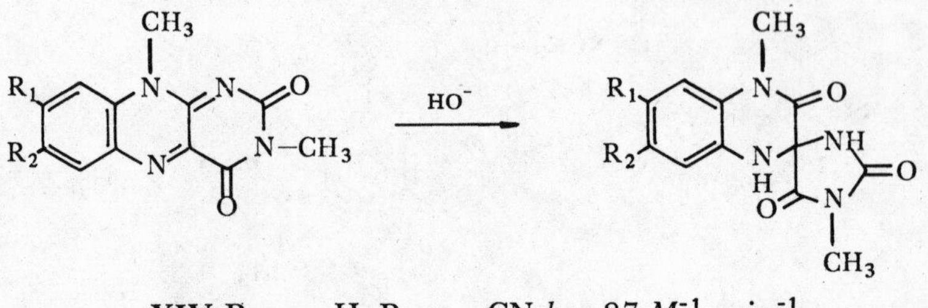


(28)

**XIV** $R_1 = -H$, $R_2 = -CN$ $k = 87\ M^{-1}\ min^{-1}$
**XV** $R_1 = -CN$, $R_2 = -H$ $k = 56\ M^{-1}\ min^{-1}$

order in hydroxide ion and is initiated at the 10a-position [78]. Alkaline hydrolysis of 1-alkylflavinium salts also provides spirohydantoins (eq. 29) [79] while 5-alkylflavinium salts exist in solution

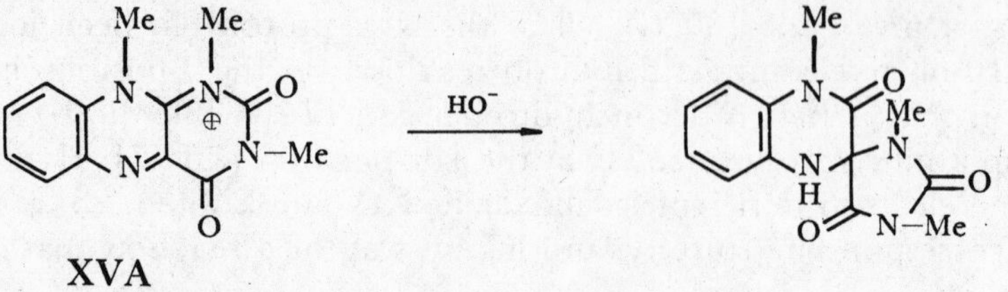


(29)

**XVA**

in equilibrium with their 4a-carbinolamines (eq. 30) [80]. N(5)-Alkyl flavinium salts undergo conversion to the corresponding radical **XVB**

$$\text{(30)}$$

and $F_{ox}$ in protic solvent (eq. 31) in a ratio of 2:1 at pH 7 [133]. Mr. Kemal [81] in the author's laboratory has established the following general-base mechanism (eq. 31 and 32):

(i.e., $FH_2$ + $CH_2O$)

For B: = $H_2O$ $k = 2.2 \times 10^{-6}$ $M^{-1}$ $sec^{-1}$      (31)
Brønsted $\beta = 0.6$

$$\frac{k[N(5)\text{-}CH_3]}{k[N(5)\text{-}CD_3]} = 10.5 \text{ for B: } = H_2O$$

$$FH_2 + 2 FMe_{ox} \rightarrow F_{ox} + 2 FMe\cdot \textbf{ XVB} \qquad (32)$$

These investigations support the 4-, 10a-, and 4a-positions as electrophilic centers susceptible to nucleophilic addition. Hemmerich and Müller [82] have argued, from the spectrum of the equilibrium product formed on reaction of MeO⁻/MeOH with a 1-alkyl-flavinium compound [83], that it is the 6-position that undergoes addition (eq. 33). [Note that F. Müller and P. Hemmerich have recently ruled out, by nmr studies, addition at the 6- or 8-position (see also p. 74). It

$$\text{(33)}$$

would now appear as though the 10a (as originally proposed [83]) and 9a-positions are favored. Hemmerich now favors position 9a [83a].].

## 2.2 Sulfite Addition: The Mechanism of 5 versus 4a-Addition and the Electrophilicity of the 8-Position

The addition of $SO_3^{2-}$ to the 5-position of flavin and 1-alkylflavinium compounds has been established by Massey and Müller [16] (eq. 34), and Müller has established [84] a similar addition of phosphines

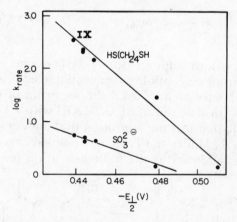

$$(34)$$

to the 5-position. In the case of the reaction of $SO_3^{2-}$, the log of the second order rate constant is a linear function of $E_{1/2}$ for the reaction $F_{ox} + 2e^- \rightarrow FH_2$ [12] (Fig. 1). Sulfite addition (eq. 34) is interesting since it represents, in its formation of a sulfamic acid of a dihydroflavin, oxidation of sulfite and reduction of flavin. Furthermore, the unusual situation of nucleophilic attack upon the nitrogen terminal of an azomethine is observed. It would be of considerable interest to ascertain if imines of highly electron-deficient carbonyl groups are

Figure 1. Linear free energy correlations with polarographic half-wave potentials of rates of reaction of isoalloxazines with 1,4-butanedithiol and sulfite ion [12].

subject to nucleophilic attack at nitrogen.

The sulfite addition reaction has served to differentiate the flavo-protein dehydrogenases and oxidases [85]. The flavin moiety of the oxidase reacts with $SO_3^{2-}$ at the 5-position, forms red (anionic) semi-quinones, and reduces $O_2$ directly to $H_2O_2$. The dehydrogenases do not react with $SO_3^{2-}$, form blue (neutral) semiquinones, and reduce $O_2$ to $O_2^{\cdot -}$. Presumably the tertiary structure of these two classes of flavoenzymes differ in that the dehydrogenases possess an acidic group in position to act as a proton donor to the free radical species and to block approach to the N(5)-C(4a) positions.

When a hydrogen occupies the 8-position of the isoalloxazine ring, the sulfite reaction is more deep seated. Hevesi and Bruice [86] have investigated the reaction of $SO_3^{2-}$ with the isoalloxazine IX, finding it to be kinetically biphasic (Scheme 4) proceeding through the

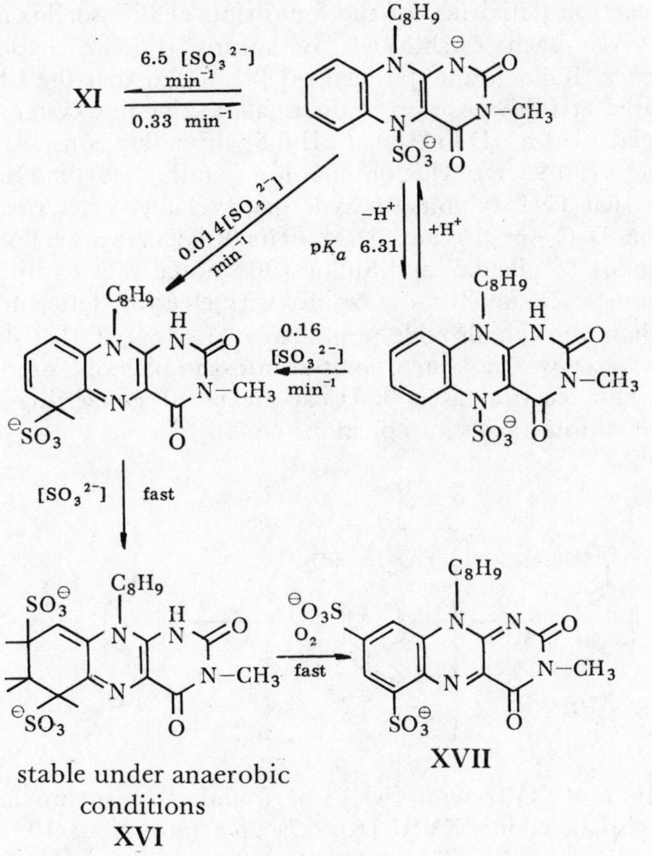

Scheme 4

formation of the 5-sulfite adduct and to yield (in the presence of $O_2$) the 6,8-disulfonic acid (**XVII**). The reaction of $SO_3^{2-}$ with the 5-adduct may occur at the 6- or the 8- position, and in either case the mechanism may involve direct nucleophilic attack accompanied by displacement or, alternatively, attack upon a nitrenium ion species (eq. 35), as established in not too dissimilar systems by Gassman and Campbell [87].

$$(35)$$

The electron deficiency of the 8-position of the isoalloxazine ring system has been established by means of two experimental approaches. Bullock and Jardetsky [88] found that the $CH_3$-group substituted at the 8-position of the isoalloxazine ring system of FMN exchanged —H for —D in $D_2O$ at pH 6.8 with a rate constant of 2.4 × $10^{-6}$ $sec^{-1}$ (90-95°C). This observation is rather startling in view of the fact that TNT exhibits no hydrogen exchange when treated with NaOD in $D_2O$ for 15 min [89]. The Russian workers Vainstein, Kukhtenko, Tomilenko, and Shilov [90] noted that hydrogen peroxide anion reacts much more rapidly with electron-deficient sulfonic acids than does hydroxide ion. They also noted that $^{18}O$ from solvent water was not incorporated into the phenolic products and offered the reaction as a practical means of converting electron-deficient sulfonic acids into phenols (eq. 36).

$$(36)$$

Reaction of **XVII** with $HO_2^-$ was found [91] to provide the 8-hydroxyisoalloxazine **XVIII** (eq. 37) at a rate about $10^2$ times as great as that for the like reaction of 2,4-dinitrobenzene sulfonate

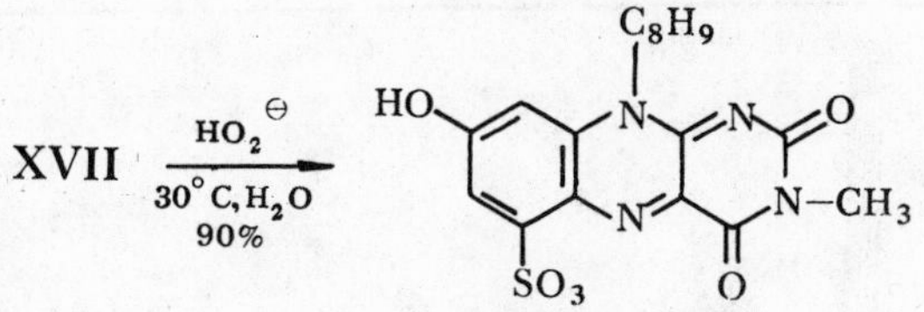

$$\text{XVII} \xrightarrow[\substack{30°C, H_2O \\ 90\%}]{HO_2^{\ominus}} \text{(structure XVIII)} \tag{37}$$

XVIII
p$K_a$ 4.45
$\lambda_{max}$ 480 nm

(compare XVIII to the naturally occurring flavin V). The p$K_a$ values of the 8-OH groups of V and XVIII are comparable to the p$K_a$ values of 2,3- and 2,4-dinitrophenol, respectively. The electron deficiency of the 8-position of the isoalloxazine ring system is bound to be of mechanistic significance, since many enzymes are attached covalently to the enzyme by way of the 8$a$-carbon (loc. cit.).

The reaction of $SO_3^{2-}$ with the isoalloxazine-6,8-disulfonate (XVII) was found to provide both 5- and 4a-sulfite adducts (XIX and XX) [92]. These could be differentiated by means of their ultraviolet

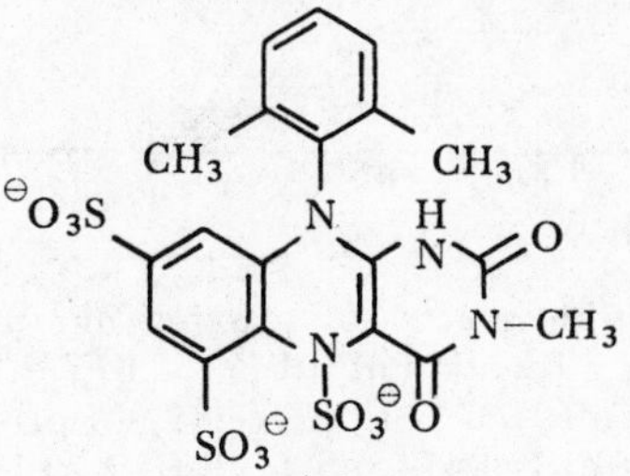

XIX

$\lambda_{max}$ 307 nm
Singlet 115 Hz 6H

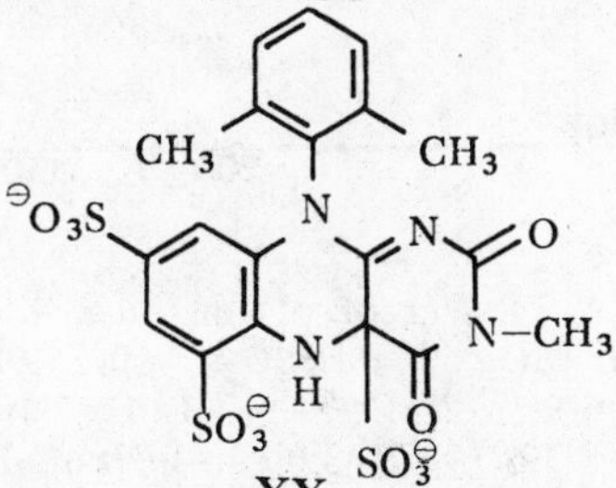

XX

$\lambda_{max}$ 304 nm shoulder
360-365 nm
Singlets 112 Hz 3H
137 Hz 3H

spectra and by nmr. Thus for XIX the 2'- and 6'-methyl groups of the 10-substituent are symmetrically located, whereas for XX the asymmetry of the molecule at the 4a-carbon causes one methyl group to be forced against the center diazine ring, resulting in a splitting of the methyl proton signals. Figure 2 shows scans of spectra of equilibrium mixtures of XVII in $SO_3^{2-}$ + $HSO_3^-$ buffers of constant buffer concentration as a function of pH. Consideration of Figure 2 reveals that with increase in pH the extent of reaction decreases (concentration of unreacted XVIII increases at $\lambda_{max}$ 445), the concentration of XIX increases, and the concentration of XX decreases.

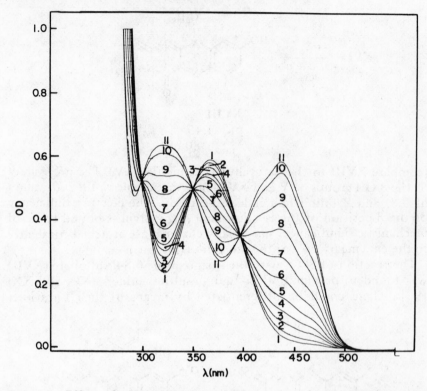

Figure 2. Effect of pH on ultraviolet-visible spectra of reaction mixtures of **XVII** in $[SO_3^{2-}]$ + $[HSO_3^-]$ = 0.149 $M$ as a function of pH ($\mu$ = 2.0, $T$ = 30°C). (1)pH=6.62, (2)pH=7.19, (3)pH=7.48, (4)pH=7.72, (5)pH=7.81, (6)pH=8.12, (7)pH=8.35, (8)pH=8.60, (9)pH=8.90, (10)pH=9.45, and (11)pH=9.79 [92].

Clearly then, the mechanisms of formation of 5- and 4a-sulfite adducts differ. Since **XX** represents the only known observable 4a-adduct formed from a simple flavin or isoalloxazine under nonphotolytic conditions, the system was exploited to determine this difference in mechanism. From a combination of kinetic and thermodynamic measurements we were able to arrive at the mechanisms of Scheme 5. The important finding in our studies is that nucleophilic addition to the 5-position does not require general catalysis, whereas 4a-addition is catalyzed by the buffer acid and, as required by microscopic reversibility, the retrograde process is catalyzed by the buffer base. General acid catalysis of 4a-addition is anticipated from examination of the change in $pK_a$ values of the N-(5) nitrogen accompanying 4a-addition (eq. 38). The N-(5) position undergoes a change in basicity of over $10^{14}$ in proceeding along the reaction coordinate.

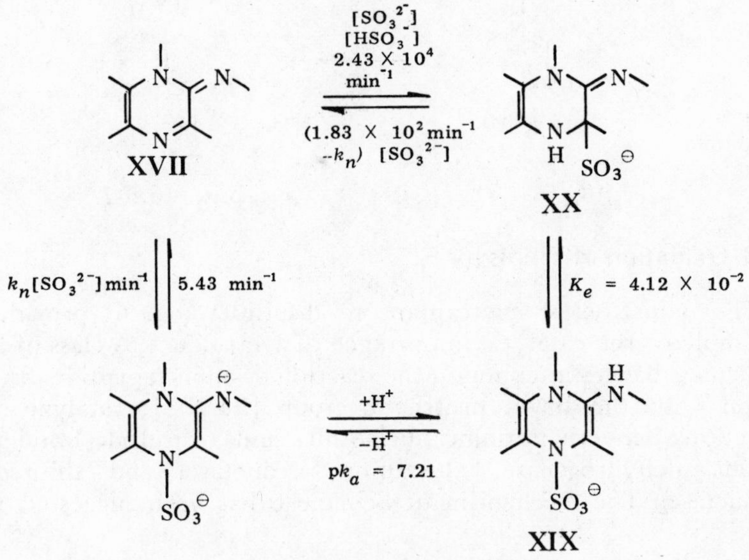
[SO3 2-]
[HSO3 -]
2.43 X 10^4
min^-1
(1.83 X 10^2 min^-1
-kn) [SO3 2-]
XVII
XX
kn[SO3 2-] min^-1
5.43 min^-1
Ke = 4.12 X 10^-2
+H+
-H+
pka = 7.21
XIX

Scheme 5

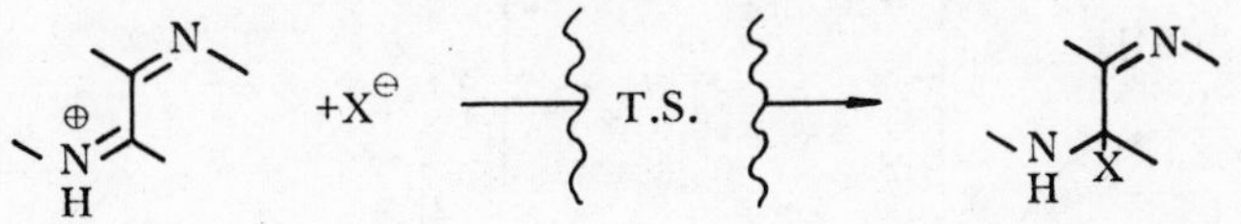
+X
T.S.

(38)

$pK_a < O$ $\qquad$ $pk_a >> pK_w$

Therefore protonation of the N(5) position on reaching the transition state (T.S.)—general-acid catalysis (eq. 39)—should greatly

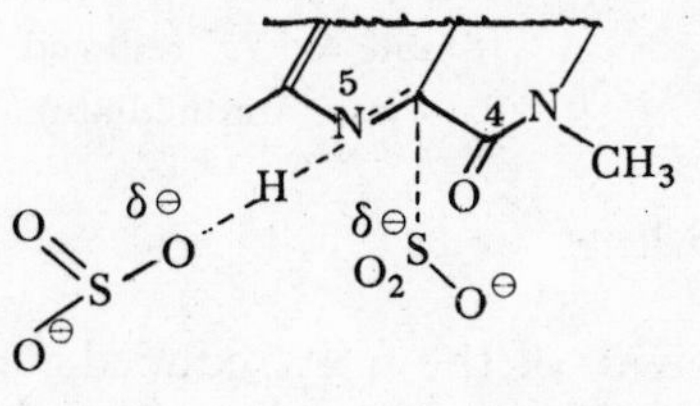
5
4
N
CH3
H
O
δ⊖
δ⊖
O2
S
O

(39)

reduce its free energy content [93] and expedite 4a-nucleophilic attack. The formation of a 5-adduct in a reaction occurring near neutrality should not be subject to general-acid catalysis since the $pK_a$ of the N—H bond at the 1-position is near 7 and the resulting anion is a stable species (eq. 40).

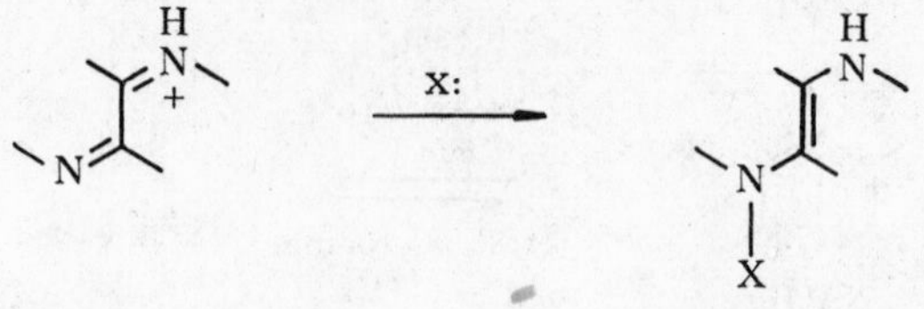
X:

$pK_a \cong 0$ $\qquad$ $pK_a = 6.5\text{-}7$ $\qquad$ (40)

## 2.3 Oxidation of Thiols by $F_{ox}$

The oxidation of mercaptans to disulfides (eq. 6) provides an example of the catalytic importance of a 4a-adduct. A class of flavo-enzymes, having in common the reaction of an intrapeptide disulfide bond with the flavin prosthetic group [94-101], catalyze redox reactions between pyridine nucleotides and a disulfide bond (lipoamide dehydrogenase, glutathione reductase, and thioredoxin reductase). The mechanism of Scheme 6 has been suggested [102,

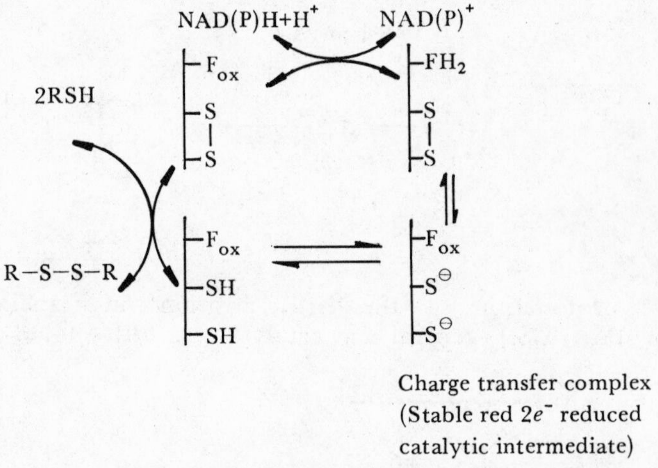
NAD(P)H+H+
NAD(P)+
Fox
FH2
2RSH
R–S–S–R
SH
SH
Charge transfer complex
(Stable red 2e⁻ reduced
catalytic intermediate)

Scheme 6

102a] on the basis of the inactivity of the fully reduced enzyme, the appearance of two extra sulfhydryl groups upon reduction of enzyme with NADH, and spectral and chemical evidence.

Three model investigations of the oxidation of mercaptans by $F_{ox}$ are pertinent [78, 103, 104]. Gascoigne and Radda [103] examined the reduction of a number of isoalloxazine derivatives by thiols and established that (*a*) the log of the relative rate constants for reduction by dihydrolipoic acid is a linear function of the flavin $E_{\frac{1}{2}}$ values (see also reference 12 and Fig. 1) and (b) the reduction by the dithiol lipoic acid is strongly catalyzed by buffer species. The employment of flavins unsubstituted at the N(3)-position and therefore ionizable in the pH-range investigated makes the kinetic data most difficult to interpret.

The reduction of 3,10-dimethyl-8-cyanoisoalloxazine (**XV**) by thiophenol and mercaptoethanol has been investigated in the author's laboratory [78]. In neither case was the reduction of this electron-deficient isoalloxazine subject to buffer catalysis. The reactions were found to proceed to completion giving dihydroisoalloxazine and disulfide in quantitative yield. The reduction with thiophenol has been studied most. At a constant pH the reaction was found to be first order in F$_{ox}$ and second order in total thiol (i.e., [RSH] + [RS$^-$]). At any constant concentration of total thiol a plot of $k_{obs}$ versus pH gave a bellshaped curve, with both the ascending and descending legs of the curve (Fig. 3) associated with the p$K_a$ of thiophenol. These results provide the kinetic expression of eq. 41 ($k_s$

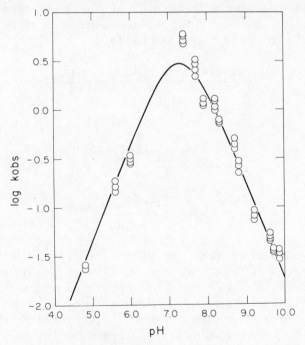

Figure 3. Dependence of the oxidation of thiolphenol (1 $\times$ 10$^{-3}$M) by **XV** (1 $\times$ 10$^{-5}$M) pH in 20% acetonitrile (30°C; $\mu$ = 1.0) [78].

$= 12 \ M^{-2} \ \text{min}^{-1}$). The rate law of eq. 41 is in accord with a general

$$- \frac{d[F_{ox}]}{dt} = k_s [C_6H_5S^{\ominus}] [C_6H_5SH] [XV] \tag{41}$$

acid-catalyzed 4a-addition followed by rate-determining nucleophilic attack of RS⁻ upon the adduct (eqs. 27 and 42). Arguments have

$$\tag{42}$$

been presented for the requirement of general-acid catalysis for 4a-addition (see Section 2.2). For the mechanism of eq. 42 there follows the kinetic expressions of eq. 43.

$$\frac{d[RSSR]}{dt} = \frac{k_1 k_2 [RS^-]^2 [AH] [F_{ox}]}{k_2 [RS^-] + k_{-1} [A^-]}$$

$$k_{-1}[A^-] \gg k_2 [RS^-] \tag{43}$$

$$V = \frac{k_1 k_2 K_{RSH} [RS^-] [RSH] [F_{ox}]}{k_{-1} K_{AH}}$$

$$k_s = \frac{k_1 k_2 K_{RSH}}{k_{-1} K_{AH}}$$

The isoalloxazine (IX) has proven useful in determining the importance of the availability of the 10a-position to nucleophilic addition. In IX the 2′- and 6′-methyl groups project in front and behind the isoalloxazine ring to provide steric hindrance to approach at the 9a- and 10a-positions (Fig. 4). As previously seen, IX does not undergo hydrolytic scission at the 10a-position. This steric inhibition to nucleophilic attack at the 10a-position is sufficient to completely

Figure 4. Stereochemical representation of 10-(2′, 6′-dimethylphenyl)-3-methyl-isoalloxazine **IX** showing steric crowding of the 10a-position by the methyl groups.

prevent the hydrolysis of **X** in alkyaline solution. This is actually quite remarkable when one considers that **X** (Scheme 2) represents an imine of urea that would be anticipated to fall apart in water. As discussed later, in no reaction examined to date is the availability of the 10a-position important to isoalloxazine or dihydroisoalloxazine redox reactions. This statement is based on the relative rates of reaction of **IX** in the particular system in question. The finding [12] that **IX** exhibits a rate constant for reaction with butane-1,4-dithiol compatible with its $E_{1/2}$ value (Fig. 1) rules out the involvement of initial addition of RS⁻ at the 10a-position. Two free radical mechanisms that would, in accord with experimental results, not require an induction period are provided in eqs. 44 and 45. For the mechanisms of eqs. 44 and 45 the rate of disulfide formation and flavin reduction would be inversely proportional to dihydroflavin concentration, since the reaction is not significantly reversible under the conditions of the experiment. This has been found not to be the case.

$$RS^- + F_{ox} \xrightleftharpoons{K_1} RS\cdot + F^{\overline{\cdot}}$$

$$F^{\overline{\cdot}} + RSH \xrightleftharpoons{K_2} FH^- + RS\cdot \qquad (44)$$

$$2RS\cdot \xrightarrow{k_3} RSSR$$

$$RSH + F_{ox} \underset{k_{-1}}{\overset{k_1}{\rightleftharpoons}} RS\cdot + HF\cdot$$

$$RS\cdot + HF\cdot \underset{k_{-2}}{\overset{k_2}{\rightleftharpoons}} RS^+ + HF^- \qquad (45)$$

$$RS^+ + RS^- \xrightarrow{k_3} RSSR$$

Nucleophilic addition of thiolate anion followed by nucleophilic attack upon the adduct by a second thiol anion was originally suggested by Coffey and Hellerman [72] for the oxidation of glutathione by 2,6-dichlorophenolindophenol (eq. 46).

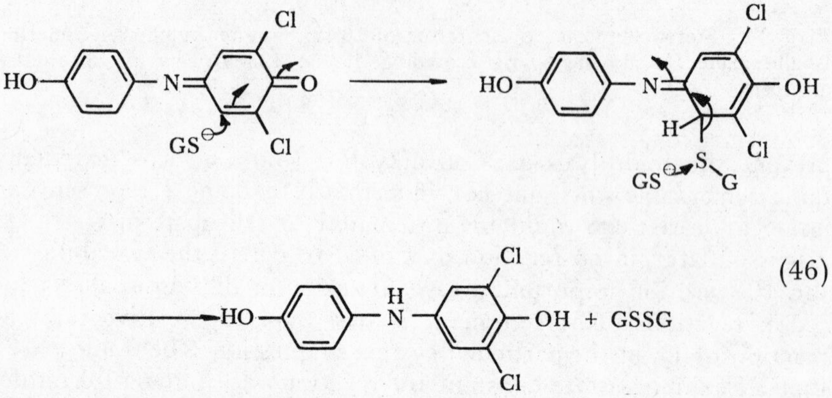

$$(46)$$

Loechler and Hollocher provide [104] the most compelling evidence for general acid catalysis in the formation of a RS⁻ flavin 4a-adduct in their investigation of the reaction of lumiflavin-3-acetic acid with dithiothreitol (DTT). Unlike the oxidation of monothiols, which do not exhibit general-acid catalysis [78, 104] but follow the rate law of eq. 41, the oxidation of DTT is first order in flavin and in dithiol and exhibits buffer catalysis up to pH 11. Loechler and Hollocker propose the reaction sequence of Scheme 7 in which the

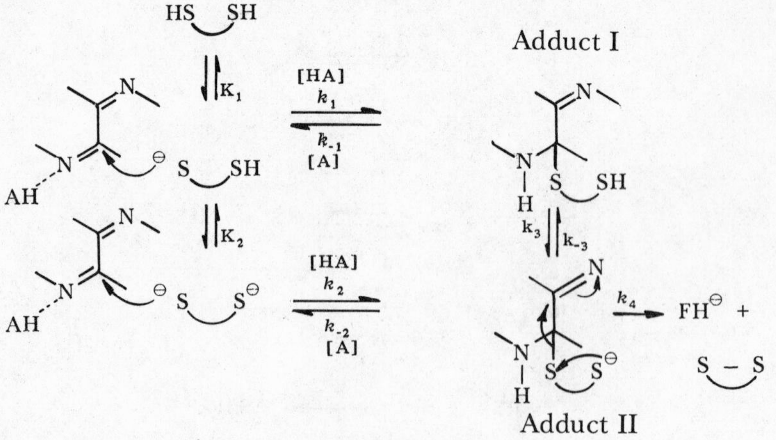

Scheme 7

attack steps 1 and 2 are rate determining, at least up to about pH 11. A pH-rate profile for the uncatalyzed reaction is consistent with Scheme 7 at least from pH 8.8 to 11.7 [$pK_1$ = 9.08 and $pK_2$ = 10.04 (by titration); $k_1$ = 4.13 $M^{-1}$ $min^{-1}$, $k_2$ = 3.10 $M^{-1}$ $min^{-1}$, $k_4 k_2/k_{-2}$ = 2.67 X $10^{-3}$ $min^{-1}$ at 25°C, $\mu$ = 1.0 with KCl]. Since the reaction of monothiols with flavin does not exhibit general catalysis but the reaction of dithiols does, it is required that if both reactions occur by the same mechanism then the monothiol and dithiol reactions must exhibit different rate-determining steps. For monothiols the rate-determining step has been suggested to be nucleophilic attack of thiol anion upon the 4a-adduct (eq. 42) [78, 104]. The reaction kinetics for the dithiol are in accord with the rate-determining step being formation of the 4a-adduct. The change in rate-determining step is due to the juxtaposing of the thiol groups in the dithiol (anchimeric, neighboring group or propinquity effect [105, 106]) which greatly increases the value of $k_2$ (eq. 42). At pH values greater than 11, a change in rate-determining step is observed in the reaction of dithiols. The new rate-determining step is no longer buffer catalyzed and has the rate law of eq. 47, which may be compared to eq. 41 and is consistent with rate-determining breakdown of inter-

$$v = k_s [HS\text{-}\wedge\!\!\!\vee\text{-}S^-] [H^+] [F_{ox}] \tag{47}$$

mediate. Loechler and Hollcher believe that breakdown becomes rate determining at high pH becuase the partitioning of Adduct II of Scheme 7 back to reactants by hydroxide ion begins to compete with the forward reaction. The situation then becomes analogous to that for monothiols and so neither system shows buffer catalysis.

## 2.4 Reaction of Carbanions with $F_{ox}$

*2.4.1* **Dehydrogenation to Form a $\overset{\diagdown}{C}=\overset{\diagup}{C}$ Bond.** Flavin-catalyzed dehydrogenations to form carbon-carbon double bonds have been studied in several model systems [19, 107]. The dehydrogenation of both dimethyl *trans*-1,2-dihydrophthalate (**XXI**) and

XXI                    XXII

dimethyl 1,4-dihydrophthalate (**XXII**) to yield dimethylphthalate is brought about by oxidized flavin. The reaction sequence, as established by Dr. Main in the author's laboratory [19], is provided in

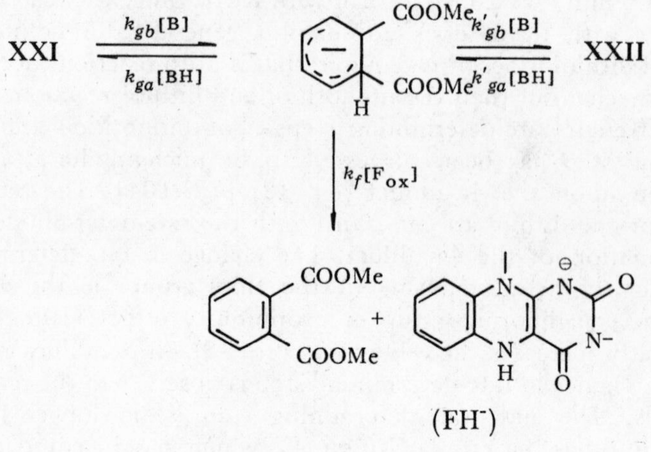

$$\text{XXI} \; \overset{k_{gb}[B]}{\underset{k_{ga}[BH]}{\rightleftarrows}} \; \overset{COOMe}{\underset{COOMe}{\bigcirc}}_{H} \; \overset{k'_{gb}[B]}{\underset{k'_{ga}[BH]}{\rightleftarrows}} \; \text{XXII}$$

$$\downarrow k_f[F_{ox}]$$

(FH⁻)

Scheme 8

Scheme 8 for which the kinetic expression of (eq. 48) pertains for

$$\frac{d[\text{FH}^-]}{dt} = \frac{k_{gb}k_f[F_{ox}][XVI][B]}{k_{ga}[BH] + k_f[F_{ox}]} \tag{48}$$

**XXI** [19]. Equation 48 predicts, as is found, that as $[F_{ox}]$ is decreased, the reaction changes from zero order (eq. 49) to first order (eq. 50) in $F_{ox}$. When B = HO⁻ the constant $k_{gb}$ (eq. 49) is

$$\frac{d[\text{FH}^-]}{dt} = k_{gb}[\text{XXI}][B] \tag{49}$$

$$\frac{d[\text{FH}^-]}{dt} = \frac{k_{gb}k_f[F_{ox}][\text{XXI}][B]}{k_{ga}[BH]} \tag{50}$$

found to equal the rate constant for isomerization of **XXI** to **XXII** (eq. 51), establishing the carbanion as a common intermediate. For

$$\text{XXI} \; \xrightarrow{216 \, [HO^-] \, (min^{-1})} \; \text{XXII} \tag{51}$$

this system $k_{gb}$ [HO⁻] is about ten times as great as $k'_{gb}$ [HO⁻] so that the reaction of eq. 51 proceeds—for all practical purposes—only in the direction written.

Knowing that the reactive species of the dihydrophthalate is the carbanion does not clarify the mechanism since the carbanion might react with flavin by a free radical mechanism or by way of a covalent adduct followed by base-catalyzed elimination (eq. 26). Whichever the case may be, it would appear that preequilibrium complexing of carbanion with oxidized flavin is involved. In the dehydrogenation of **XXI** there is no apparent relationship between log $k_{rate}$ and $E_{1/2}$. The log of the partition coefficient of the intermediate carbanion ($k_f/k_{ga}$ of Scheme 8) between phthalate ester and carbon acid starting material is, however, a linear function of the log of the complexing constants of oxidized flavins with either tryptophan or β-resorcylic acid [19]. The derived overall kinetic pathway (eq. 52) requires that

$$\begin{array}{c} -\overset{|}{\underset{|}{C}}-H \\ -\overset{|}{\underset{|}{C}}-H \end{array} \xrightleftharpoons[k_{ga}[BH]]{k_{gb}[B]} \begin{array}{c} -\overset{|}{\underset{|}{C}}{}^{\ominus} \\ -\overset{|}{\underset{|}{C}}-H \end{array} \rightleftharpoons \text{Complex} \longrightarrow \begin{array}{c} \diagdown\diagup \\ C \\ \parallel \\ C \\ \diagup\diagdown \end{array} + FH_2 \qquad (52)$$

if the complex is formed spontaneously between carbanion and F$_{ox}$, its conversion to products is spontaneous. Since complex formation should occur in an uncatalyzed diffusion-controlled process, addition of carbanion to F$_{ox}$ in an intracomplex reaction is not anticipated to be at the 4a-position but could be at the 5-position. Of course, carbanion addition may not be involved, since a radical process is quite feasible because of the anticipated stability of **XXIII**. Involvement of the 10a-position may be ruled out since the rate of reaction

**XXIII**

of **IX** with carbanion is as anticipated on the basis of its equilibrium constant for complexation of tryptophan or β-resorcylic acid. Unfortunately, meaningful results that would implicate or rule out the formation of an adduct from the complex (eq. 53) were not obtained.

Complex $\longrightarrow$ $\longrightarrow$ Products    (53)

### 2.4.2 Oxidation of α-Hydroxyketones, α-Hydroxyaldehydes, α-Hydroxy Esters and α-Hydroxyamides.

The oxidation of α-hydroxycarbonyl compounds by $F_{ox}$, as studied in our laboratory, has been found to be either first order in [HO$^-$] and [$F_{ox}$] or general-base catalyzed and zero order in [$F_{ox}$], depending on the particular substrate. These results dictate the sequence of eq. 54 and the general rate law of eq. 55 [108]. Further support for involvement of

$$v = \frac{k_{gb}k_f[B][F_{ox}][CH]}{k_{ga}[BH] + k_f[F_{ox}]} \qquad (55)$$

the 1,2-enediolate anion as an intermediate derives from the observation that the zero order rate constants for oxidation of benzoin and furoin are similar regardless if the oxidizing agent is $O_2$, 2,6-dichlorophenolindophenol or lumiflavin [109], and that there is a C–H(D) isotope effect of $k^H/k^D \cong 7$ for the base catalyzed oxidation of benzoin by lumiflavin. Again one may consider a covalent addition

of carbanion (eq. 56) or oxyanion (eq. 57) or alternatively a free radical mechanism forming, as an intermediate, the highly stabilized

$$(57)$$

semidione radical [110, 111]. $Fl^{\bullet}$ pair which does not dissociate prior to product formation since the rate of $F_{ox}$ production does not equal ½ that for benzoin ionization (eq. 58). On the basis of the

$$2 \; F \cdot \xrightarrow{k_3} F_{ox} + FH^-$$

$$(58)$$

$$\frac{d[FH_2]}{dt} = \frac{k_1[B][CH]}{2} \quad \text{where } k_2[F_{ox}] \gg k_{-1}[BH]$$

forgoing kinetic results, the mechanism intuitively derived by Brown and Hamilton [112] (eq. 59) for the flavin oxidation of α-ketol

$$(59)$$

structures is ruled out. Relative rate constants for α-ketol oxidation by lumiflavin are provided in eq. 60 [108, 109].

$$
\begin{array}{ccccccc}
C_6H_5 & CH_3 & H & & NH_2 & NH_2 \\
| & | & | & CN & | & | \\
C{=}O & C{=}O & C{=}O & | & C{=}O & C{=}O \\
| & | & | & HC{-}OH & | & | \\
HC{-}OH & HC{-}OH & HC{-}OH & | & HC{-}OH & HC{-}OH \\
| & | & | & H & | & | \\
C_6H_5 & CH_3 & H & & C_6H_5 & CH_3
\end{array} \tag{60}
$$

$$
k_{rel} \cong 1 \quad 7 \times 10^{-2} \quad 2 \times 10^{-2} \quad 2 \times 10^{-4} \quad 1 \times 10^{-5} \quad 1 \times 10^{-5}
$$

*2.4.3* **Reaction with Nitroalkane Anions.** The reaction of nitro-alkane anions with $F_{ox}$ represents an additional class of carbanion reactions. In the reaction of D-amino acid oxidase with nitroalkanes to produce aldehyde and nitrite ion, the nitroalkane anion species was shown to serve as substrate and the mechanism of eq. 61 was proposed [113]. Nucleophilic attack of carbanion at the 5-position

of $F_{ox}$ is certainly feasible. Though flavins per se are unable to undergo the overall reaction of eq. 61, we have found that the electron-deficient isoalloxazine **XV** reacts with nitromethane, nitro-ethane, and 2-nitropropane in aqueous solution to yield nitrite ion, the corresponding carbonyl compound, and dihydroflavin. The reactions with nitromethane and 2-nitropropane were found to be first order in **XV** and first order in nitroalkane anion and were not catalyzed by acid or base species. The mechanism proposed by Porter, Voet, and Bright for the reaction of D-amino acid oxidase (eq. 61) with nitroalkane anion would appear quite reasonable for the nonenzymatic reaction. The kinetically equivalent mechanism of general-acid-catalyzed 4a-addition of nitroalkane anion followed by rate-determining hydroxide attack (eq. 62) must be considered. Since

$$
\underset{\underset{NO_2}{|}}{-\overset{|}{C}}(-) \;+\; F_{ox} \;\underset{k_{gb}[B]}{\overset{k_{ga}[BH]}{\rightleftharpoons}}\; \text{4a-adduct} \;\xrightarrow{k_n[HO^-]}\; \text{Product}
$$

(62)

$$
k_{obs} = \frac{k_{ga}K_w}{k_{gb}K_{BH}} \; [-\overset{\ominus}{\underset{|}{C}}-NO_2]
$$

nitroalkane anions are ambident nucleophiles, undergoing both C and O alkylation [110, 111], the $k_n[HO^-]$ step of eq. 62 may be considered to be represented by eqs. 63 to 65. The mechanism of eq. 63

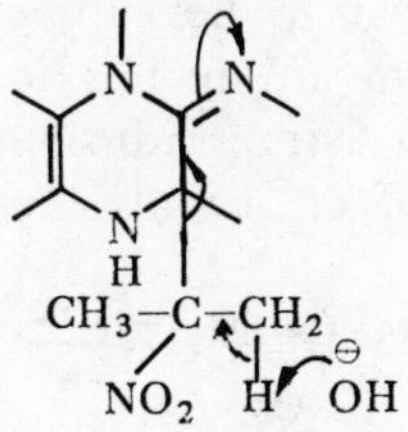

(63)

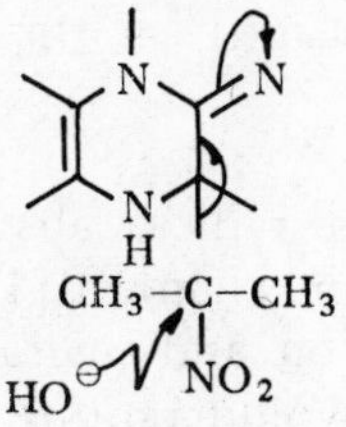

(64)

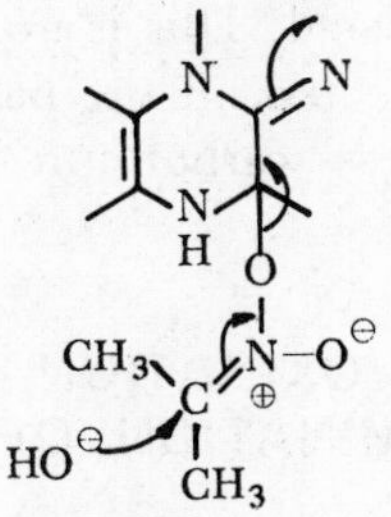

(65)

may be discounted on the basis that the base-catalyzed proton abstraction and elimination should possess an inordinately great free energy of activation. The mechanism of eq. 64 should be prohibited on the basis of steric considerations and the fact that 4a-alkyl substituted flavins do not exhibit nucleophilic displacement by $HO^-$ [114]. The process given by eq. 65 must be considered possible. If the mechanism of nitroalkane anion oxidation by $F_{ox}$ follows the path of eqs. 62 and 65, it should be possible for a buffer base species to replace hydroxide ion (eqs. 66 and 67). As stated previously, these

$$-\overset{|}{\underset{|}{\underset{NO_2}{C}}}(-) \ + \ F_{ox} \ \underset{k_{gb}[B]}{\overset{k_{ga}[BH]}{\rightleftarrows}} \ \text{4a-adduct} \xrightarrow{k_n[B]} \text{Product} \qquad (66)$$

$$k_{obs} = \frac{k_{ga}k_n \, [BH] \, [-\overset{\ominus}{\underset{|}{C}} - NO_2]}{(k_{gb} + k_n)} \qquad (67)$$

reactions do not appear to be subject to general catalysis. The C-alkylation of nitroalkane anions has been known since the early work of Kornblum to be of a free radical nature [111, 115-117]. Therefore the mechanism of eq. 68 also requires consideration. Here the

$$F_{ox} \ + \ CH_3 - \overset{\ominus}{\underset{\underset{NO_2}{|}}{C}} - CH_3 \ \underset{-H^+}{\overset{+H^+}{\rightleftarrows}} \ F^{\overline{\cdot}} \ + \ CH_3 - \overset{\cdot}{\underset{\underset{NO_2}{|}}{C}} - CH_3$$

$$\xrightarrow{\hspace{1cm}} FH^{\ominus} \ + \ CH_3 - \underset{\underset{NO_2}{|}}{C} = CH_2 \quad \text{etc.}$$

$$(68)$$

initially formed flavin radical abstracts H· from the nitroalkane radical. An additional mechanism [117a] would involve nucleophilic attack at the 4a-position and migration to the 5-position. A mechanism of this nature would require general-acid catalized attack of carbanion and would thus appear kinetically as general-base catalyzed addition of carbon acid. This is not the case. Further, a 4a → 5 shift would not be expected on the basis of the inability of the migrating group to possess carbonium ion stability (see following section).

## 3 MECHANISMS OF OXIDATION, MIGRATION, AND BASE-CATALYZED ELIMINATION OF 4a- AND 5-ALKYL $FH_2$ ADDUCTS

These reactions have been investigated by Dr. Clerin [114] in our laboratory in order to provide a mechanistic basis to judge the reasonableness of the occurrence of carbon bonded 4a- and 5-adducts as intermediates in some flavin-catalyzed reactions. The hypothetical reaction of eq. 69 may be considered. In the reactions of eq. 69, 3-

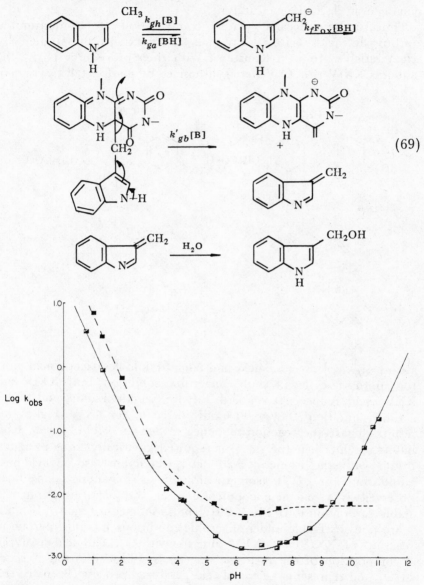

(69)

Figure 5. Log $k_{obs}$ versus pH rate profile for solvolysis of 4a-InH (−) and 4a-InMe (−). Solvent: $H_2O$ (aerobic), 30°C, $\mu = 0.2$. Rate constants in $min^{-1}$ [114].

methylindole is oxidized to 3-hydroxymethylindole with the accompanied reduction of flavin. Though the direct oxidation of 3-methylindole by flavin is not an established reaction per se, it has analogies in flavoenzyme catalysis.

Photocatalytic decarboxylative condensation of $\alpha,\beta$-unsaturated carboxylic acids with flavins has been extensively studied by Hemmerich's group in Konstanz. With their procedures [118] the adducts **XXIV** and **XXV** were synthesized by photocatalytic conden-

XXIV                                          XXV

sation of indole-3-acetic acid and $N$-methylindole-3-acetic acid with the appropriate flavin. Under anaerobic conditions both **XXIV** and **XXV** yield reduced flavin and 3-hydroxymethylindole, and under aerobic conditions they yield oxidized flavin. For **XXIV** (Fig. 5) our results dictate the reaction sequence of Scheme 9 [114]. The acid-base 4a-adduct pair ($pK_a$ 4.4) is required kinetically to be in equilibrium with the isomeric 5-adducts, which are also in acid-base equilibrium ($pK_a$ 8.1). Isomerization of 4a → 5-adducts have been observed as a general phenomenon when the migrating group is a stable carbonium ion [119, 120]. The spontaneous terms of Scheme 9 are not observable under anaerobic conditions and must pertain to reaction of **XXIV** and **XXV** directly with $O_2$. The acid-catalyzed decomposition of the 4a-adduct species is enhanced in the presence of $O_2$ and represents a general-acid catalyzed process (Br$\phi$nsted $\alpha \cong$ $-0.5$) in accord with eq. 70. The base-catalyzed reactions indicated

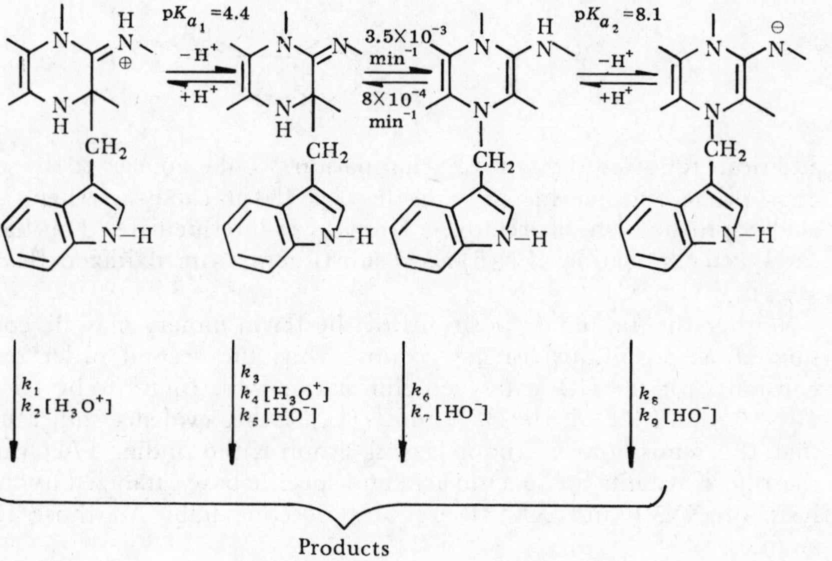

Products

Scheme 9

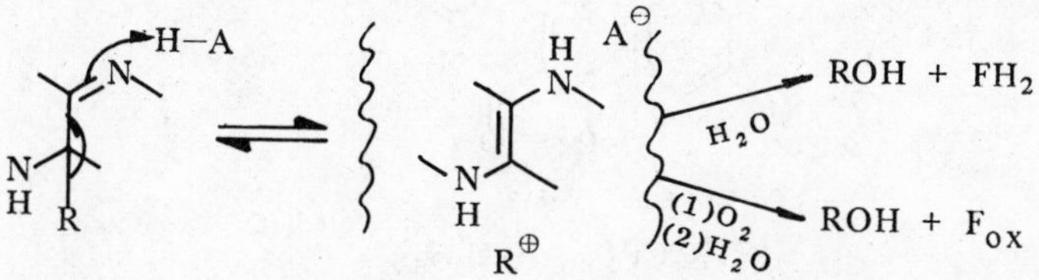

$$(70)$$

by $k_5$, $k_7$, and $k_9$ are observed only with **XXIV** and not with **XXV**. Also, these rate constants are not affected by the admittance of O$_2$. In addition, $k_5$, at least, represents a general-base catalyzed reaction (Brønsted $\beta \cong +0.5$). These findings are in accord only with an E2 elimination reaction (eq. 71). From these results nucleophilic

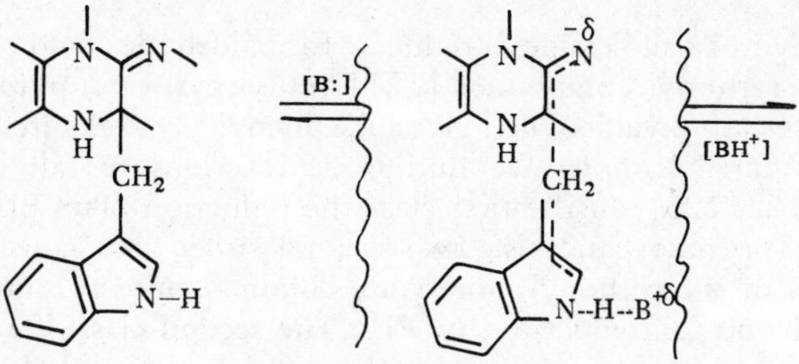

$$(71)$$

addition followed by an E2 elimination would appear to be an appropriate mechanism to consider for flavin catalysis when, an acidic proton can be removed from a 4a-intermediate. For base catalyzed elimination of N(5)-alkyl substituent from oxidized flavin see eq. 31, p. 21.

Neither the 4a- nor 5- positions of the flavin moiety may be considered as propitious leaving groups. Thus the second order rate constants for the HO⁻ catalyzed elimination were found to be $1.1 \times 10^2$ $(k_5)$ and $2.5 \times 10^1$ $M^{-1}$ $min^{-1}$ $(k_9)$. Other evidence indicating that the 5-position is a poor leaving group is the finding [74] that the rate constants for specific-acid and specific-base-catalyzed hydrolysis of **XXVI** and **XXVII** are at best comparable to those for anilides.

**XXVI**     R=C₆H₅
**XXVII**    R=CH₃

## 4   REACTIONS OF FH₂ WITH ORGANIC SUBSTRATES

### 4.1   Carbonyl group Reduction by FH₂

1,5-Dihydroisoalloxazines reduce formaldehyde and electron deficient carbonyl compounds [121] (ethyl pyruvate, pyruvic acid, pyridine-4-carboxyaldehyde, glyoxal, phenyl glyoxal, etc.) to the corresponding alcohols. We find these reactions to fall into two classes [122, 123]. In the first class the reduction takes place without acid or base catalysis. Two second order rate constants for reduction of the carbonyl moiety are obtained, one for reduction by FH₂ and one for reduction by FH⁻. The second class of substrates requires hydrogen ion catalysis for the reaction of both FH₂ and FH⁻ with substrate. The most studied example is the reduction of formaldehyde. Chloral oxidizes both FH$_2$ and FH$^-$ to F$_{ox}$ in a non-acid catalyzed reaction which is first order in chloral and both dihydro-flavin species. 2,2,2, - Trichloroethanol is not, as has been reported [121], a product of this reaction. Thus chloral does not undergo a 2e$^-$ reduction by FH$_2$ or FH$^-$. Formation of a radical [i.e., Cl$_3$ĊH-(OH)] in a le$^-$ transfer seems likely (eq. 72). Chloride ion has been

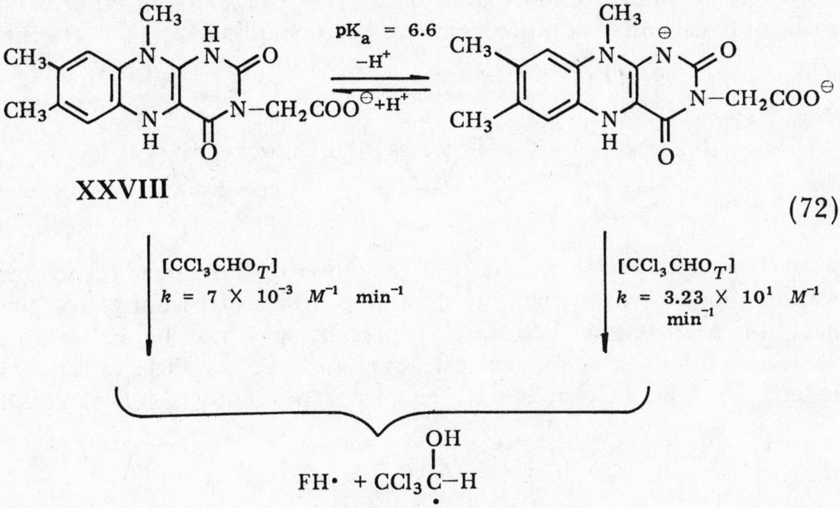

**XXVIII**

$$pK_a = 6.6$$

(72)

$[CCl_3CHO_T]$

$k = 7 \times 10^{-3}$ $M^{-1}$ min$^{-1}$

$[CCl_3CHO_T]$

$k = 3.23 \times 10^{1}$ $M^{-1}$ min$^{-1}$

$$FH\cdot + CCl_3\overset{OH}{\underset{\cdot}{C}}-H$$

then

$$2\ FH\cdot \xrightarrow{fast} FH_2 + F_{ox}$$

found as a product of the reaction.

Covalent intermediates that may be considered in carbonyl group reduction include **XXIX**, **XXX** and **XXXI**. The formation of **XXIX** as proposed by Brown and Hamilton [112] is by inspection judged

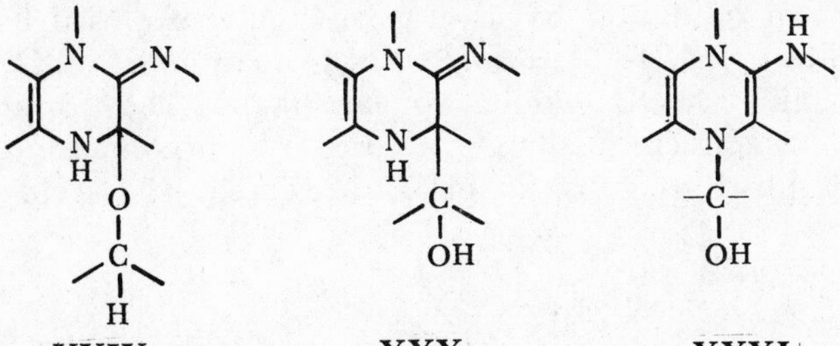

**XXIX**          **XXX**          **XXXI**

unfavorable in comparison to **XXX** since it would require nucleo-philic attack of an enamine anion (i.e., FH$^-$) upon the carbonyl oxygen (in preference to the carbonyl carbon **XXX**) with proton donation to the carbonyl carbon. Nucleophilic attack upon a carbonyl oxygen in preference to carbonyl carbon must be considered, at present, a rare phenomenon. Perhaps the only known examples of nucleophilic attack by carbon involve the addition of triphenylmethyllithium and *tert*-butylmagnesium chloride to tetracyclone, as reported by Dimroth and Laufenberg [124] (eq. 73). In eq. 73 the carbonyl carbon is quite hindered, precluding attack at this

$$(73)$$

position, while attack on oxygen is favored by the great resonance stabilization of the carbanion. Enamine attack at carbonyl oxygen has not been established but at present may not be ruled out. Thus enamines do reduce tetracyclone [125, 126] as shown in eq. 74. The mechanism of eq. 74 is not known, but it could

$$(74)$$

represent nucleophilic attack on oxygen. In any event, kinetic and isotope studies have shown that **XXIX** is not an intermediate in $\alpha$-ketol oxidation by $F_{ox}$ as originally proposed by Brown and Hamilton [112]. *A priori* the formation of **XXX** or the 5-carbinolamine **XXXI** would appear likely. The formation of **XXX** has a direct analogy in the 5-hydroxymethylation of uracil and substituted uracils [127, 128] (eq. 75). Acid catalysis

$$(75)$$

of the formation of **XXX** is reasonable, as is actually observed in the reduction of formaldehyde (eq. 76), Decomposition

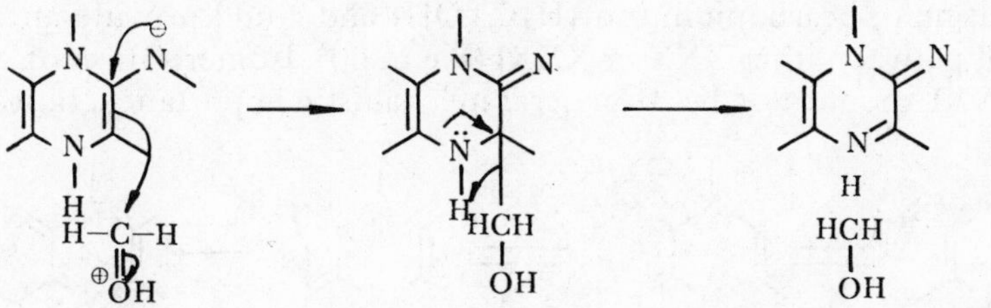

(76)

of **XXX** to products would necessitate a more or less concerted reaction where departure of $(-)CH_2OH$ would be assisted by proton transfer to this species by the incipient proton attached to N(5) (eq. 76). Proton transfer would be required since the free energy of formation of $(-)CH_2OH$ at neutrality would be prohibitive (i.e., $> \Delta G^{\ddagger}$) and there would be little driving force for the elimination of this species since the p$K_a$ of the nitrogen at the 5-position of 4a-adducts is 4.0 or less [114]. The overall sequence of events in eq. 76 involves the transfer of an electron pair plus a proton and would, in the absence of the identification of an intermediate, appear as a hydride transfer reaction. In the retrograde direction the mechanism of eq. 76 requires formation of **XXX** from $CH_3OH$. A two-electron process would require a concerted reaction in which the N(5) position of $F_{ox}$ (p$K_a \cong 0$) abstracts a C–H proton from methanol. This is most unlikely. Formation of **XXX** by attack of $(-)CH_2OH$ is even less likely. These are most important considerations since $F_{ox} + CH_3OH$ yields $FH_2 + CH_2O$ [138]! We believe that a more likely process is the free radical mechanism of eq. 78. The covalent adduct **XXXI** was postulated by Blankenhorn et al. [121] to be an intermediate in formaldehyde reduction as shown in eq. 77. That carbinolamine is

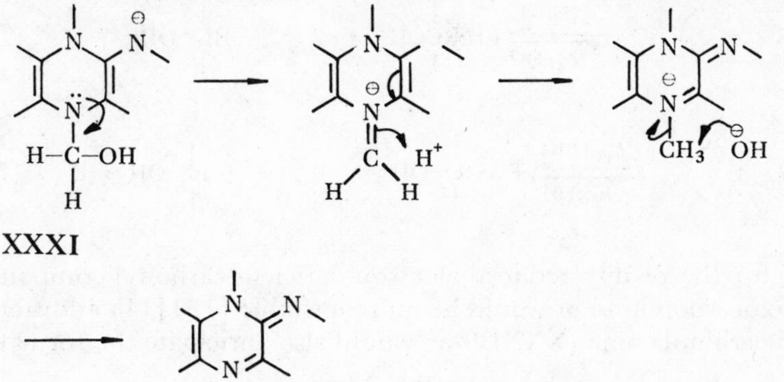

**XXXI**

(77)

formed [130] can be shown by its trapping with cyanobrohydride to yield 1,5-dihydro-5-methyl flavin [123]. In addition, isomerization of **XXX** ⇌ **XXXI** should be facile because of the stability of the migrating oxocarbonium ion ($H_2C^+=OH$) and could provide an additional path to either **XXX** or **XXXI** (see p. 00). Isomerization of **XXX** ⇌ **XXXI** should not be of general mechanistic importance, however,

$$\text{H--C--H} \qquad \text{HCH} \qquad \text{HCH} \qquad \text{CH}_3 \qquad (78a)$$
$$\underset{O}{\Vert} \qquad \underset{O-}{\vert} \qquad \underset{OH}{\vert} \qquad \underset{OH}{\vert}$$

$$\underset{/}{\overset{\backslash}{}}C=O \;\overset{+H^+}{\underset{-H^+}{\rightleftharpoons}}\; \underset{/}{\overset{\backslash}{}}C=\overset{\oplus}{O}H \;\overset{FH^-}{\rightleftharpoons}\; FH\cdot\cdot\overset{\vert}{\underset{\vert}{C}}-OH \;\rightarrow\; F_{ox} + H-\overset{\vert}{\underset{\vert}{C}}-OH \qquad (78b)$$

$$FH^- + \underset{/}{\overset{\backslash}{}}C=O \;\rightleftharpoons\; FH\cdot\cdot\overset{\vert}{\underset{\vert}{C}}-O^\ominus \;\overset{+H^+}{\underset{-H^+}{\rightleftharpoons}}\; FH\cdot\cdot\overset{\vert}{\underset{\vert}{C}}-OH \;\rightarrow\; F_{ox}$$
$$\qquad\qquad\qquad + \; (-)\overset{\vert}{\underset{\vert}{C}}-OH + H^+ \qquad\qquad\qquad (78c)$$

$$\underset{/}{\overset{\backslash}{}}C=O \;\overset{+H^+}{\underset{-H^+}{\rightleftharpoons}}\; \underset{/}{\overset{\backslash}{}}C=\overset{\oplus}{O}H \;\overset{FH}{\rightleftharpoons}\; FH\cdot\cdot\overset{\vert}{\underset{\vert}{C}}-OH \;\rightarrow\; F_{ox}$$
$$\qquad\qquad\qquad + \; (-)\overset{\vert}{\underset{\vert}{C}}-OH + H^+ \qquad\qquad\qquad (78d)$$

$$FH^\ominus + \underset{/}{\overset{\backslash}{}}C=O \;\overset{k_{ga}[BH]}{\underset{k_{gb}[B]}{\rightleftharpoons}}\; FH\cdot\cdot\overset{\vert}{\underset{\vert}{C}}OH \;\rightarrow\; F_{ox} + H\overset{\vert}{\underset{\vert}{C}}-OH \qquad (78e)$$

$$FH^- + \underset{/}{\overset{\backslash}{}}C=O \;\overset{k_{ga}[BH]}{\underset{k_{gb}[B]}{\rightleftharpoons}}\; FH\cdot\cdot\overset{\vert}{\underset{\vert}{C}}-OH \;\rightarrow\; F_{ox} + (-)\overset{\vert}{\underset{\vert}{C}}-OH + H^+ \qquad (78f)$$

since for the readily reduced electron-deficient carbonyl compounds the oxocarbonium ion would be quite unstable [131]. In addition to the 5-carbinolamine (**XXXI**) one would also anticipate the formation of the 1-carbinolamine (**XXXIA**) by way of reaction of undissociated 1,5-dihydroflavin. This expectation is based on the structural analogy

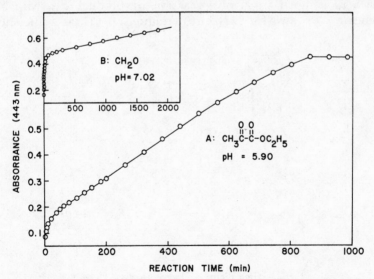

of 1,5-dihydroisoalloxazines and uracils and the knowledge that uracils form the hydroxymethyl adducts **XXXIB** [127]. An important observation is that the 1,5-dihydro-5-methyl flavin (**XXXII**) reduces chloral, ethyl pyruvate, and so forth [132]. It is impossible to write a mechanism involving a 5-adduct (as **XXXI**) as an intermediate for these reactions. This finding suggests one of two alternatives: (1) with non-N(5)-substituted dihydroflavins the mechanism of reduction of carbonyl compounds is different than

Figure 6. Time dependence of absorbance of lumiflavin-3-acetic acid (443 nm): (*A*) [ethyl pyruvate] = 0.10 *M*; [FH$_2$] = 6 × 10$^{-5}$ *M*; (*B*) [CH$_2$O] = 0.10 *M*; [FH$_2$] = 7 × 10$^{-5}$ *M* [30°C, solvent: H$_2$O (10 vol of CH$_3$OH), $\mu$ = 1.0 with KCl, N$_2$ atmosphere]. The formalin solution was prepared by thermal depolymerization of paraformaldehyde and trapping of gaseous formaldehyde in doubly glass distilled water [122].

that with **XXXII** or (2) mechanisms involving covalent adducts at the 5-position are incorrect. The latter is eminently more reasonable. In addition, dihydroflavin readily reduces benzyl [108] and for this substrate it is impossible to build a Stuart and Briegleb model for a N(1)-carbinolamine adduct. Thus, of the covalent two-electron transfer mechanisms considered (**XXIX, XXX, XXXI**), none appears to remain in contention. The carbinolamie **XXXI** is not formed along the reaction path for reduction.

Reaction of formaldehyde, ethyl pyruvate, or pyruvic acid with $FH_2$ results in a biphasic reaction (Fig. 6). The first phase is acid catalyzed, first order in $FH_2$ and carbonyl compound, and involves competitive formation of $F_{ox}$ (35% yield at all pH values) and **XXXI** and (probably) **XXXIA** in equilibrium. Saturation by formaldehyde occurs for the $FH_2$ but not the $FH^{\ominus}$ species. Thus N(1)-alkyl uracils form N(3)-hydroxymethyl uracils through the N(3)-protic but not the N(3)-anion species [127]. In the second, and much slower phase, **XXXI** is converted to $F_{ox}$ in a reaction that is zero order in both carbonyl compound and dihydroflavin. Correct analogue simulation of the time dependence of $F_{ox}$ formation is obtained if the second phase is assumed to involve reaction of product $F_{ox}$ with **XXXI** (Scheme 10 shown for $FH_2$ but pertaining to $FH^-$ as well). The rate

*Phase I*

$$FH_2 + CH_2O \underset{k_2}{\overset{k_2 a_H}{\rightleftarrows}} XXXI$$

$$k_1 a_H \downarrow$$

$$F_{ox} + CH_3OH$$

*Phase II*

$$F_{ox} + XXXI \xrightarrow{slow} FH_2 + F_{ox}-CH_2OH$$

$$F_{ox}-CH_2OH \xrightarrow{fast} F_{ox} + CH_2O$$

$FH_2$ then reenters the reaction of Phase I.

Scheme 10

constants for acid catalyzed reduction of $CH_2O$ by $FH_2$ and $FH^-$ (Scheme 10, $k_1$) are readily availably (for $FH_2$, $1.2 \times 10^{-3}$ $M^{-2}$ $sec^{-1}$;

for FH⁻, 1.0 M⁻² sec⁻¹).

For carbonyl group reduction by dihydroflavin, the free radical mechanisms of eq. 78 are proposed by the author. The following considerations reveal that the 1,5-dihydroisoalloxazine ring system should be prone to radical processes. The central dihydropyrazine ring of the 1,5-dihydroisoalloxazine structure contains 8π available electrons and is therefore *potentially* antiaromatic. From molecular orbital calculations, Streitwieser [132a] predicts that the dihydro-pyrazine ring structure should be thermodynamically destabilized because the antiaromaticity of the 8π-system requires the placing of an electron pair into an antibonding orbital. However, the degree of antiaromaticity depends on the ability of the nitrogen lone pairs to interact as part of a π-system. Molecular orbital sigma-pi approxima-tions [132b] predict that the normally bent configurations of 1,4-dihydropyrazines, oxazines, and dioxadiene preclude antiaromatic character. For this reason the butterfly configuration of 1,5-dihydro-isoalloxazines is not expected to be antiaromatic in the ground state. Flattening of the 1,5-dihydroisoalloxazine ring would be required in order to provide significant antiaromaticity. Thus it has been suggest-ed that some degree of antiaromaticity of 1,4-dialkyl-1,4-dihydro-pyrazines may be conferred by substituents on the dihydropyrazine ring [132c]. If flattening should occur removal of an electron from the antibonding orbital of the 8π-system would provide a thermo-dynamically more stable nonantiaromatic flavin radical. The dihydro-flavin moiety could reach a degree of planarity in the transition state accompanying a one electron transfer thus facilitating such a transfer. In any explanation of the ease of one electron transfer from FH$_2$ and FH⊖ one must consider the degree of lone-pair splitting of the electrons on the N(5) and N(10)-nitrogens. The lone pair-lone pair interaction in systems N̈−C−C−N̈ are such that splitting occur and this may be relieved by radical formation [132d]. Here planarity is not desired. A suitable model is triethylenediamine, which exhibits a reversible one electron oxidation to yield a relatively long lived (seconds) radical cation [132e]. Planar or butterfly in conformation, dihydroflavins should readily undergo one electron transfer reactions. This has not previously been recognized.

The reactions of eq. 78 have been written for the anionic FH⁻ species but pertain to FH$_2$ as well. In eq. 78a, electron transfer from dihydroisoalloxazine to aldehyde yields the radical anion that upon protonation provides the generally lesser reducing ·CH$_2$OH species, that then abstracts a hydrogen atom from the flavin radical cation to yield products. Reaction 78b and 78d to 78f differ from 78a in that formation of $\cdot\overset{\mid}{\underset{\mid}{C}}-O^-$ is obviated by either specific-acid or general-acid catalysis. In eq. 78e and 78f proton and electron transfer is concerted. The likelihood of these mechanisms is real since the radical anions ($\cdot\overset{\mid}{\underset{\mid}{C}}-O^-$) are less stable than the neutral radical species ($\cdot\overset{\mid}{\underset{\mid}{C}}-OH$) [133]. The overall sequence of events in eqs. 78a, 78b, and 78e involves, as does eq. 76, the transfer of two electrons plus a proton from reduced flavin to substrate and in the absence of identification of intermediates would appear as a hydride transfer reaction. Electron abstraction (eqs. 78c, 78d, and 78f) is to be expected in competition with H$\cdot$ abstraction from FH$\cdot$ by $\cdot\overset{\mid}{\underset{\mid}{C}}-OH$. This process would yield initially the carbanion species [i.e., $(-)\overset{\frown}{\underset{\mid}{C}}OH$] and would not be favored in the absence of resonance or considerable inductive stabilization of the carbanion electron pair. In instances where resonance or inductive stabilization is possible, it is anticipated that H$\cdot$ and $e^-$ abstraction should be competitive (see p. 65). In favor of the mechanisms of eq. 78 is the stability of the radical species generated from the carbonyl [133, 134] and flavin [5] components and the fact that radical abstraction of H$\cdot$ is a ubiquitous process, being particularly favorable when the H$\cdot$ donor is bonded to the conjugated nitrogen [135]. Indeed the H$\cdot$ donor ability of conjugated secondary amines is the basis of their industrial use as antioxidants [136]. Radical abstraction of H$\cdot$ from a conjugated secondary amine radical (HN$\overset{\frown}{\cdot}$ +)—as is the flavin radical species—is a highly exergonic reaction and could approach a diffusion-controlled process.

The following considerations lead to a resonable reaction coordinate diagram for a radical process of $CH_2O$ reduction by $FH_2$. The value of $E_0'$ for the half-cell of eq. 79 is sufficiently positive to

$$\cdot CH_2O^{\ominus} \rightleftharpoons CH_2O + e^- \qquad E_0' = +1.4860 \qquad (79)$$

prohibit the formation of $\cdot CH_2O^{\ominus}$ in the dihydroflavin reduction of formaldehyde. In addition, the anion $(-)CH_2OH$ has too high a free energy to be a product. This dictates that the mechanisms of eq. 78b and 78e must be considered for formaldehyde reduction. Of a series of oxy-acid buffers investigated, only the most acidic provided what might be consisdered as feasible general-acid catalysis [123]. It is likely that the specific-acid-catalyzed reaction is most correct so that almost complete transfer of a proton to the carbonyl group is accompanied by little, if any, electron transfer in the critical transition state (**XXXIC**). With the assumption of the reasonable p$K_a$ of -10 for CH$_2$O (p$K_a$ of CH$_3$CHO$\overset{+}{\text{H}}$ is 10.2 [137], the $\Delta F^0$ of forma-

$$\overset{-\delta}{\text{FH}\cdot}-\cdot---\overset{+\delta}{\underset{/}{\overset{\backslash}{\text{C}}}}=\text{O}--\text{H}----\overset{+\delta}{\text{O}}\overset{\text{H}}{\underset{\text{H}}{\diagup}}$$

## XXXIC

tion of H$_2$C=$\overset{+}{\text{O}}$H from CH$_2$O may be computed at any pH value. The value of $\Delta F^0$ of formation of the radical pair   FH·· CH$_2$O—OH   as well as product may be computed from existing (eq. 80 and 81) [5, 133, 134] electrochemical data [123, 138].

$$\text{FH}_2 \;\rightleftharpoons\; \text{·FH} + e^- + \text{H}^+ \qquad E_0' = -0.188$$

$$\text{FH}^- \;\rightleftharpoons\; \text{·FH} + e^- \qquad\qquad E_0' = -0.146 \qquad (80)$$

$$\text{FH}^- \;\rightleftharpoons\; \text{·F}^- + e^- + \text{H}^+ \qquad E_0' = -0.0295$$

$$\text{·CH}_2\text{OH} \;\rightleftharpoons\; \text{CH}_2\text{O} + e^- + \text{H}^+ \qquad E_0' = +0.357 \qquad (81)$$

If we assume acid-catalyzed formation of the neutral radical moiety ·$\overset{|}{\text{C}}$—OH is rate determining in all cases, and with a knowledge of the $\Delta F^0$ of formation of intermediates and products as well as $\Delta F\ddagger$, we may construct a reaction coordinate diagram for formaldehyde (Fig. 7). It is important to note that no corrections have been estimated or applied to the calculated free energy levels for what must be free radical pair intermediates to allow for their stabilization by cage effects, charge transfer, and so on. Consequently, the calculated $\Delta F^0$ levels represent maximum values. For Figure 7 the values of $k_1$ come directly from $K_a (=k_{-1}/k_1)$ and the knowledge that acid dissociations in the highly exothermic direction are diffusion controlled ($k_{-1} = 10^{10}$ sec$^{-1}$). The rate constants for electron transfer from $F_{\text{red}}$ to $\underset{/}{\overset{\backslash}{\text{C}}}$=$\overset{+}{\text{O}}$H ($k_2 = 10^9$-$10^{10}$ $M^{-1}$ sec$^{-1}$) are in agreement with other rate constants for thermodynamically favorable electron transfer reactions (flavin included) [9, 10, 133, 139]. The highly exergonic abstraction of H· from FH· and FH$_2^{+}$ should be particularly favorable ($10^9$ sec$^{-1}$) since the flavin moieties are not only conjugated secondary amines but, most importantly, radical cations

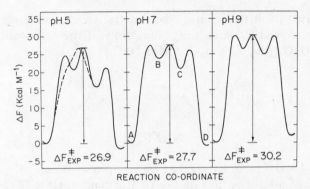

Figure 7. Computed reaction coordinate diagrams at pH 5, 7, and 9 for the reduction of $CH_2O$ by 1,5-dihydroflavin species: $(A)$ $CH_2O$ + total dihydroflavin; $(B)$ $CH_2OH$ + total dihydroflavin; $(C)$ $\cdot CH_2OH$ + total flavin radical species; and $(D)$ $CH_3OH$ + $F_{ox}$. The values of $\Delta F^{\ddagger}$ refer to the experimentally determined rate of reaction. The dashed line in the pH 5 figure pertains to a reaction in which state $C$ is reached by means of concerted $H^+$ and $1e^-$ transfer.

(i.e., $\overset{+}{:}NH$). For those carbonyl compounds that do not require acid catalysis in their reduction by $FH_2$, it is anticipated that the formation of the initial $\cdot CH_2O^-$ species is rate determining. Progress in this area requires reliable values of the $\Delta G^0$ of formation and $pK_a$ values for radical species (see p. 63 for general discussion).

As stated previously, a number of carbonyl compounds (chloral, ethyl pyruvate, quinone, ninhydrin, etc.) are reduced by **XXXII** to yield its free radical (eq. 82) [132]. The formation of **XXXIII** from

$$ \text{\textbf{XXXII}} \xrightarrow{\phantom{xx}\,C=O\,\phantom{xx}} \text{\textbf{XXXIII}} \qquad (82) $$

**XXXII** may occur (1) by $1e^-$ transfer from **XXXII** to carbonyl compound or (2) by $2e^-$ reduction of carbonyl compound followed by a comproportionation reaction (eq. 1) between unreacted **XXXII** and formed oxidized 5-methyl flavin. A choice between $1e^-$ and $2e^-$ transfer may be made on the basis that $(a)$ 1 mole of ninhydrin yields 1 mole of **XXXIII**, $(b)$ the production of **XXXIII** on reaction of benzoquinone with **XXXII** is associated with a rate constant that exceeds that for comproportionation, and $(c)$ reaction of chloral

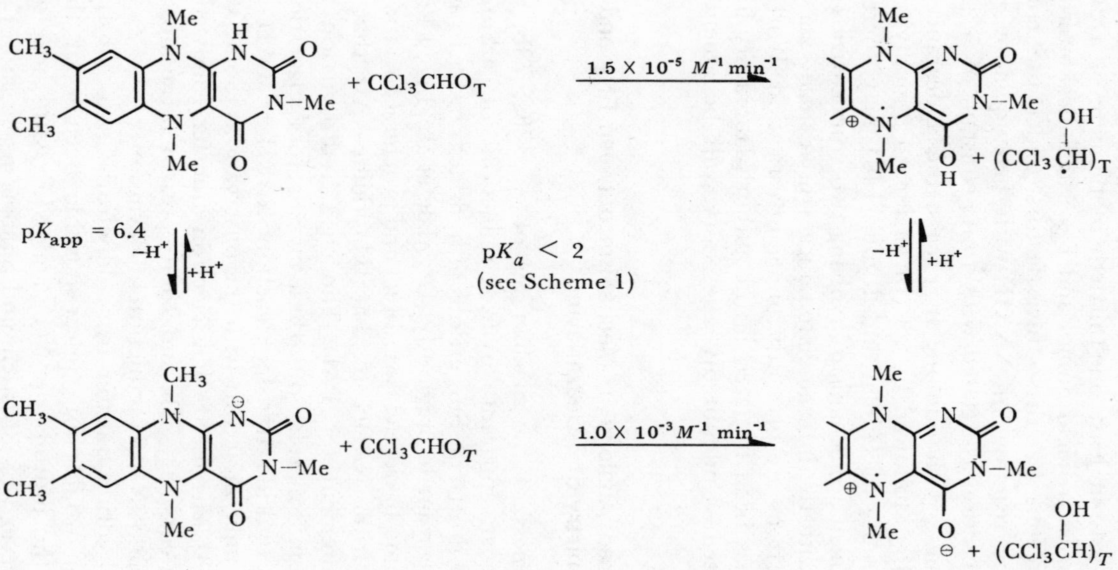

$$(83)$$

Me
$N$
Me
$N$—Me
$\overset{+}{\cdot}$
O
H
$O + (CCl_3\overset{\cdot}{C}H)_T$
OH

$\Big\updownarrow$ $\begin{array}{c} -H^+ \\ +H^+ \end{array}$

Me
$N$
Me
$N$—Me
$\overset{+}{\cdot}$
O
$\ominus$
$O + (CCl_3\overset{\cdot}{C}H)_T$
OH

**XXXIII**
$\lambda_{max}$ 502;580

$\dfrac{1.5 \times 10^{-5} \, M^{-1} \, min^{-1}}{}$

$\dfrac{1.0 \times 10^{-3} \, M^{-1} \, min^{-1}}{}$

$pK_a < 2$
(see Scheme 1)

CH₃
$N$
CH₃
$N$—Me
Me
$N$
H
$N$—Me
O
O
$+ CCl_3CHO_T$

$pK_{app} = 6.4$ $\begin{array}{c} +H^+ \\ -H^+ \end{array}$

CH₃
$N$
CH₃
$N$—Me
CH₃
$N$
$N$—Me
O
$\ominus$
O
$+ CCl_3CHO_T$

**XXXII**

55

with **XXXII** does not yield the $2e^-$ reduction product $2',2',2'$-tri-chloroethanol. The reactions of eq. 82 occur under the same conditions as reduction of the substrates by $FH_2$. The reaction of chloral with **XXXII** is satisfied by the sequence of eq. 83. Gibian and Rynd [140] found the reaction of $FH_2$ with quinone occurred in aqueous solution at a rate at least one-thirtieth of the rate of a diffusion-controlled process to yield quinol and $F_{ox}$. No radical intermediates could be detected by these investigators under the conditions employed. The reaction of **XXXII** with benzoquinone provides **XXXIII** in a reaction associated with a rate constant of $\geq 10^8$ sec$^{-1}$ at all pH values. With quinone in excess, the appearance of the radical **XXXIII** is followed by its conversion to 5-methyl oxidized flavin. The results of Gibian and Rynd [140] may therefore be explained through a mechanism involving two consecutive $1e^-$ transfer reactions with the intermediate radical pair at steady state. These particular reactions are presently under investigation. Work in this general area has been initiated in the author's laboratory but is far from completed. Undoubtedly the story will be amended and extended.

## 4.2 A Proposed Analogy in Mechanism between FH₂ and NADH. The Chemistry of 5-Deazaflavins

The correctness of the mechanisms of eqs. 78a, 78b, and 78e would require an apparent "hydride reduction" of carbonyl compound by dihydroflavin. Suggestions of hydride transfer to and from flavins have been made previously. Drysdale et al. [141, 142] found with cytochrome b₅ reductase a direct and prochiral specific transfer of hydrogen from NADH to 3-acetylpyridine-NAD that had an absolute requirement for FAD. Though a hydrogen ion trapping mechanism cannot be ruled out, these investigators' suggestion of the stereospecific hydride transfer reaction NADH → FAD → NAD analog is certainly in accord with the observations. Some years ago Kosower [143] suggested that electron transfer followed by H· transfer (as in eqs. 78a, 78b, and 78e) represented the most reasonable mechanism of NADH reduction of flavins.

Proof of a hydride transfer from dihydroflavin to substrate in a model system would not appear possible becuase of the rapid exchange of the protons on N(1) and N(5) with those of protic solvents. However Drs. Brüstlein and Shinkai in the author's laboratory have shown that 5-deazisoalloxazine (**XXXIV**) undergoes the hydride transfer reactions of eqs. 84 and 85 in $D_2O$ [122, 144]. One

$$\text{XXXIV} \qquad \qquad R = N\text{-peopyl,} \qquad\qquad (84)$$
$$R = AD$$

XXXV

$$\text{XXXV} + \qquad\qquad\qquad \xrightarrow{D_2O} \qquad \text{XXXIV} + \qquad\qquad (85)$$

might argue that the replacement of the nitrogen at the 5-position by carbon provides an analog of NADH rather than a flavin [73]. In the view of the author, this may not be an important distinction, since it is likely that both dihydropyridines and dihydroflavins employ a common reductive mechanism (eq. 78). Dihydroflavin, in comparison to NADH, is mechanistically a more versatile reducing agent, being able, as we have seen in thiol oxidation and so forth, to operate also by covalent addition reactions and radical mechanisms not involving an overall "hydride" transfer (see above). It has been established in two laboratories [145, 146] on the basis of the non-identity of kinetic deuterium isotope effects and the corresponding product isotope effects, that dihydropyridine reductions involve the formation of a metastable intermediate (eq. 86). This metastable intermediate cannot be a charge transfer complex because its slow rate of formation can be calculated to be close to that for the overall reaction. In addition, the intermediate must have a $t_{1/2}$ sufficiently long to differentiate between —H and —D. The author proposes that an intimate radical pair be considered to represent this intermediate.

$$v = \frac{k_1 k_2 \, [\text{NADH}]\,[-\overset{\overset{\displaystyle O}{\|}}{C}-]}{k_{-1} + k_2} \qquad k_{\text{rate}} = \frac{k_1 k_2}{k_{-1} + k_2}$$

$$\frac{k_{\text{rate}}^{H}}{k_{\text{rate}}^{D}} = \text{Kinetic isotope effect}$$

$$\frac{k_2^{H}}{k_2^{D}} = \text{Product isotope effect}$$

The reduction of thiobenzophenone by $N$-propyldihydronicotin-amide has been shown to provide the thiol anion radical **XXXVI** [147, 148] identifiable by means of epr. Structure **XXXVI** was

**XXXVI**

suggested to be formed in a radical pair since the reaction was not inhibited by free radical trapping agents. There is no evidence that the detected radical is a kinetically competent intermediate at present. These investigations have led to the suggestion or implicit statement that dihydropyridine reduction follows the course of $1e^- + 1e^- + H^+$ transfer [145-148]. However the mechanism of $1e^-$ transfer followed by H· transfer is preferred by this author for $N$-alkyl-dihydronicotinamide ($RPyrH_2$) reductions since (1) it explains best the lack of mixing of hydrogen nuclei with protic solvent and (2) it obviates the necessity of formation of the high free energy contain-ing protonated pyridinium ion that would be required in a $1e^- + 1e^- +$ intramolecular $1H^+$ transfer reaction ($RPyH_2^{2+}$)—i.e., $RPyrH_2 + S \rightleftharpoons RPyRH_2^+ \cdot S^- \rightleftharpoons RPyrH_2^{2+} S^- \rightleftharpoons RPyrH\,HS$.

When the log of the second order rate constants for reduction of a series of isoalloxazines by either NADH or $N$-($n$-propyl)-dihydroni-cotinamide [NPNH] are plotted versus the $E_{1/2}$ values for the iso-alloxazines, no linear free energy relationship is found between rate and potential. On the other hand, there is a relationship

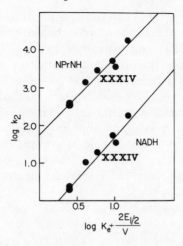

Figure 8. Plot of log $(K_e + 2E_{1/2}/\text{volts})$ versus the log of the second order rate constant for reduction of a series of isoalloxazines by $N(n\text{-propyl})$-dihydronicotinamide [NPrNH] and NADH.

between the log $k_{\text{rate}}$ and the log of the equilibrium constants for complexing of isoalloxazines by either tryptophan or $\beta$-resorcylic acid [12]. The fit of the points to the plot is greatly improved if a four-parameter linear free energy relationship is employed that takes into account the electron density of the flavin $(E_{1/2})$ and complexing abiltiy (log $K_e$) (see Fig. 8). We have interpreted these observations as indicating preequilibrium complexing of dihydropyridine and isoalloxazine moieties prior to electron transfer [12]. The suggestion of preequilibrium complexing has been strengthened by the finding [149] of saturation kinetics in the reduction of lumiflavin by NADH. That reaction does not involve the 9-, 9a-, 10-, and 10a-positions of the isoalloxazine ring system is attested to by the fact that the point for **IX** does not deviate from the linear free energy plot relating rate to complexing ability and potential. Thus in both the dehydrogenation of dimethyl *trans*-1,2-dihydrophthalate (p. 35) and the oxidation of dihydronicotinamide, preequilibrium complexing of reactants appears to be involved, and in both instances the region of the isoalloxazine nucleus surrounding the 10-position is unimportant (Fig. 4). Complexes of $FADH_2$ and $NAD(P)^+$ have been observed (by spectral means) with glutathione reductase [97], $p$-hydroxybenzoate hydroxylase [102], NADH-rubredoxin reductase [150], and melilotate hydroxylase [151]. The complexes with $p$-hydroxybenzoate hydroxylase and melilotate hydroxylase have been shown to arise along the reaction paths of catalysis. In two very interesting studies, Blankenhorn has shown charge transfer in both intermolecular [152] and intramolecular [153] complexing of dihydropyridine and flavin. Complexing of planar molecules in aqueous solution is a most common phenomenon. The major contributing factors to complex stability are minimization of interfacial tension and interfacial solvent organization [154-156]. In water electronic factors such as dipole-dipole interaction and charge transfer, are likely to play a minor role in complex stability. The occurrence of charge transfer within the complex, once formed, may or may not portend to eventual complete electron transfer with the generation of radical species.

Regarding the analogies between **XXXIV** and flavins, it should be pointed out that (1) the uv-visible spectrum of **XXXIV** closely resembles that of oxidized flavin, (2) the rate constants for reduction of **XXXIV** by dihydropyridines fit the same linear free energy relationship as do the isoalloxazines (Fig. 8), (3) **XXXIV** adds $SO_3^{2-}$ at the 5-position in a reversible manner [144] as does oxidized flavin, and (4) **XXXV** is converted (albeit much more slowly than 1,5-dihydroflavin) by $O_2$ oxidation to **XXXIV** [144, 157]. Song has argued, from SDN, FOD, and $\pi_{r,r}$ calculations, that since the 5-position is the most electrophilic position of the isoalloxazine and 5-deazaisoalloxazine ring systems, the reaction patterns of both should be the same [158]. In a comparison of 1,5-dihydroflavin and its N(5)-deaza analog the following considerations must be taken into account: First, the 5-position of the dihydroflavin is much more electronegative. In addition, the 1,4-dihydropyrazine ring of the dihydroflavin possesses *potential* antiaromaticity (p. 51) and N(5)-N(10) orbital splitting. These features are not shared by the N(5)-dihydrodeazaflavine. Antiaromaticity should not, however, be evident in the ground state because of the inability to realize an $8\pi$-planar system in the butterfly conformation. It is possible that in the transition states for some reactions the dihydroflavin moiety possesses a planar or partially planar conformation, with the resultant promotion of two electrons to an antibonding orbital. This feature would be relized in model systems through preequilibrium complexation of reactants or in the enzymatic reaction by restriction of the ground state conformation. If this were so the induced antiaromaticity should be of kinetic significance because of the enhanced reactivity of the electrons in the antibonding orbital. A common feature of the N(5)-deaza analog and dihydroflavin is that the 4a-position is an enamine terminus. This position should be more nucleophilic in the case of the N(5)-deaza analog.

Most importantly, 5-deazaflavins serve as cofactors for at least five enzymatic reactions. The NADH:FMN oxidoreductase from *Beneckea harvyei* has been shown by Fisher and Walsh [159] to stereospecifically mediate the catalysis of transfer of hydrogen from NADH to 5-deazariboflavin. Jorns and Hersh [160] have established stereospecific hydrogen transfer from L-glutamate to *N*-methyl glutamate synthetase bound 5-deaza-FMN. In the presence of methyl amine, stereospecific hydrogen transfer from the 1,5-dihydro-5-deaza-FMN moiety occurs to yield *N*-methyl-L-glutamate. In addition, replacement of the normal flavin cofactor of succinate dehydrogenase by its 5-deaza analog actually provides enhanced activity [161]. Studies to determine if transfer of hydrogen from succinate to the 5-position of the deazaflavin occurs have not been carried out. The great importance of doing so is obvious. The ability to substitute an N(5)-deazaflavin cofactor for its biological counterpart in enzymatic systems and obtain hydrogen transfer to the 5-position provides evidence that the substrate moiety in the productive ES-complex resides in the vicinity of the 4a- and 5-positions. This is a most important observation. In addition, employment of the N(5)-deaza analog allows the elucidation of the stereochemistry at the active site, since, as in the case of the A and B sides of dihydronicotinamides, the hydrogens at the 5-position are prochiral. *M. smegmatis* L-lactic acid oxidase [161a] and Old Yellow Enzyme [161b] utilize the 5-deaza analog of the respective flavin cofactors.

## 4.3  Carbonium Ion Reduction by FH$_2$

The reduction of 5,10-methylene-H$_4$-folate (**XXXVIA**) to 5-methyl-H$_4$-folate (**XXXVIB**) by **XXXVIA** reductase has been established to occur by way of the sequence of eq. 87 [162, 163]. As shown by the borohydride reduction of **XXXVIA** to **XXXVIB** [164], the

$$\text{NADH} \searrow \quad \nearrow \text{FADH}_2 \searrow \quad \nearrow \text{XXXVIB}$$
$$\text{NAD} \nearrow \quad \searrow \text{FAD} \nearrow \quad \searrow \text{XXXVIA}$$

$$(87)$$

former exists to some extent as the iminium cation (eq. 88) [165].

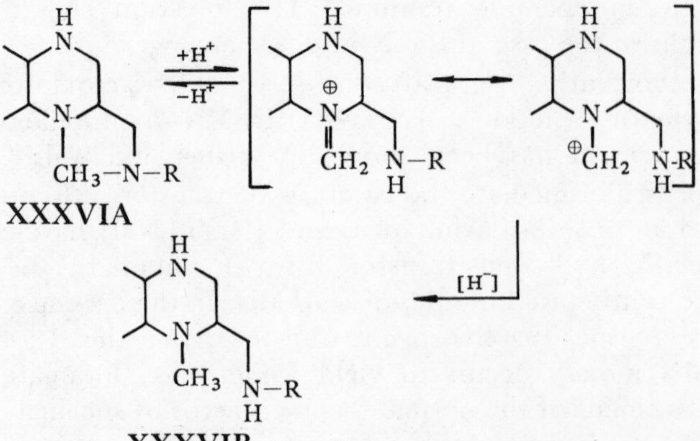

**XXXVIA**

**XXXVIB**

$$(88)$$

The stable carbocation malachite green is known to be reduced by $N$-benzyldihydronicotinamide by means of hydrogen transfer [166]. The foregoing discussion might lead one to anticipate the mechanism of eq 89. A like mechanism would be considered as a possible route

$$-\overset{|}{\underset{|}{C}}{}^{+} + NADH \rightleftharpoons \{ -\overset{|}{\underset{|}{C}}{}\cdot \ \ NADH \overset{\cdot}{:} \} \rightarrow -\overset{|}{\underset{|}{C}}H + NAD^{\oplus} \quad (89)$$

to carbonium ion reduction by $FH_2$. At pH 5 malachite green is reduced by 1,5-dihydrolumiflavin-3-acetic acid [132] in a reaction clearly second order in carbonium ion (eq. 90, Fig. 9). These results suggest a free radical mechanism (e.g., eq. 91) in which the rate of

$$\frac{-d[FH_2]}{dt} = 2.6 \times 10^5 [FH_2][-\overset{|}{\underset{|}{C}}{}^{+}]^2 (sec^{-1}) \quad (90)$$

the reaction of $FH_2 \overset{\cdot}{:}$ with $-\overset{|}{\underset{|}{C}}{}^{+}$ exceeds that for H· transfer from

$$FH_2 + -\overset{|}{\underset{|}{C}}{}^{\oplus} \rightleftharpoons FH_2 \overset{\cdot}{:} + -\overset{|}{\underset{|}{C}}{}\cdot$$

$$FH_2 \overset{\cdot}{:} + -\overset{|}{\underset{|}{C}}{}^{\oplus} \rightleftharpoons F_{ox} + -\overset{|}{\underset{|}{C}}{}\cdot \quad (91)$$

$$2 -\overset{|}{\underset{|}{C}}{}\cdot \rightarrow Product$$

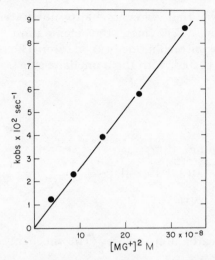

Figure 9.  Plot of the square of the concentration of malachite green ([MG]$^2$) versus the observed first order rate constant ($k_{obs}$) for reduction of the dye by 1,5-dihydrolumiflavin-3-acetic acid (pH = 5, 30$^\circ$C, solvent H$_2$O [132].

FH$^{+}$ to $-\overset{|}{\underset{|}{C}}\cdot$. Attempts to examine the reduction of **XXXVIA** [132] or the more stable analog **XXXVIC** [167] by FH$_2$, were not possible because of competing hydrolysis of substrate.

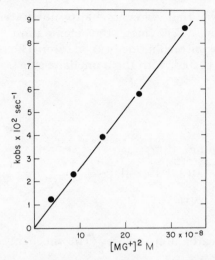

BF$_4^-$

## 4.4   Reversible Reactions

We have seen that anionic species serve as the immediate substrates for F$_{ox}$ reduction to FH$_2$ in the oxidation of thiols (p. 28), in the dehydrogenation to form a carbon-carbon double bond (p. 33) [19], in the oxidation of α-hydroxy carbonyl compounds to α-diketones (p. 36; reference 168), and in the oxidation of nitroalkanes to nitrite ion and carbonyl compounds (p. 38). Carbanions may also serve as the intermediate procursors in flavoenzyme catalyzed oxidations. Lactic acid oxidase and amino acid oxidases carry out the reactions of eq. 9 and 10. Abeles and co-workers [169-

171] have shown that when α-chloro-substituted substrates are employed for these enzymes, the elimination of chloride ion competes with the normal mode of oxidation. The reaction sequence of eq. 92 was suggested, with the immediate substrate for $F_{ox}$ oxida-

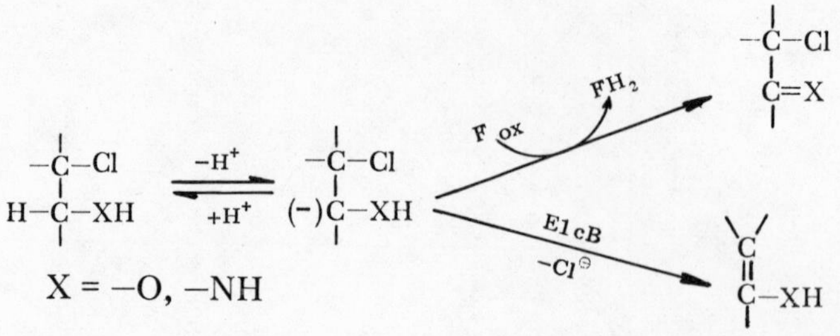

(92)

$$X = -O, -NH$$

tion being the α-carbanion species. (The mechanism of eq. 93 has

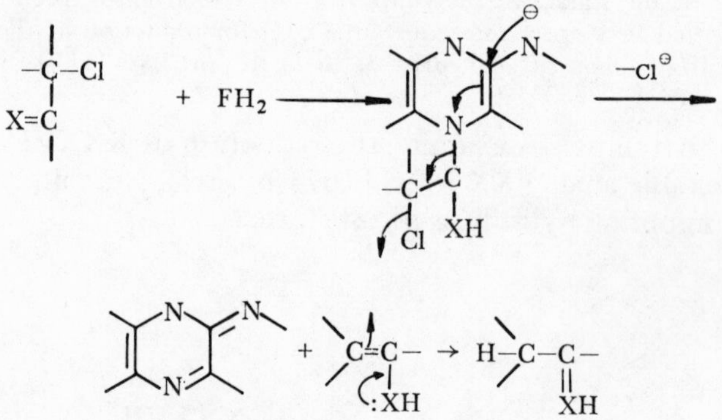

(93)

been suggested as an alternative explanation for Cl⁻ production [172]). Alternately Cl· expulsion from Cl–Ċ–CH(XH)– could be involved. Although adducts occur along the reaction path for thiol (models) and nitroalkane anion (enzymatic) oxidation, there is no direct kinetic or spectral evidence to implicate adducts as intermediates in other carbanion oxidations or in the oxidation of alcohols.

The differentiation between a radical and a covalent mechanism could depend on the most favored mode of disposition of an intermediate radical pair that could dissociate to radical products, undergo an additional $1e^-$ or H· transfer reaction, or collapse to a covalent product. Both thiol and nitroalkane ions are prone to form radicals, and it has been shown for the latter that nucleophilic substitution on carbon is radical in nature [111, 115-157] (alkylation of nitroalkane anions serve as a model for the present proposal). The considerable carbon affinity of sulfur dictates that the radical pair (i.e., FH··SR) collapse to provide an adduct. Further, dependent on the standard free energy of formation of the intermediate radical pair from covalent intermediate the latter may (1) be a dead-end side product, (2) re-enter the radical path by dissociation to the radical pair, or (3) undergo further covalent transformation to provide substrate and F$_{ox}$ [for an example of (2) and (3), see eq. 93a]. A like mechanism may be

$$FH^{\ominus} + Y \rightleftharpoons T.S. \rightleftharpoons FH\cdot\cdot Y \rightleftharpoons T.S. \rightleftharpoons F_{ox} + Y^{\ominus} + H^{+}$$

$$FH-Y \rightleftharpoons F_{ox} + Y(-) + H^{\ominus} \qquad (93a)$$

involved for the reaction of O$_2$ with dihydroflavin (see p. 73). Covalent intermediates may, of course, arise by direct nucleophilic attack particularly when the nucleophile is highly polarized as is SO$_3^{\ominus}$ and RS$^-$.

In the reduction of simple carbonyl compounds by FH$_2$, a strong case has been made for free radical mechanisms leading to alcohol (eq. 94) or carbanion (eq. 95). The various mechanisms for free

$$FH_2 + \overset{\diagdown}{\underset{\diagup}{C}}=O \rightarrow F_{ox} + H\overset{|}{\underset{|}{C}}-OH \qquad (94)$$

$$F_{ox} + (-)\overset{|}{\underset{|}{C}}-OH \rightarrow FH^- + \overset{\diagdown}{\underset{\diagup}{C}}=O \qquad (95)$$

radical carbonyl group (eqs. 78a-78f) or imine reduction may be best examined by means of the type of diagram and equations of Scheme 11.

*Step 1*

*Step 2*    (ignoring the involvement of acid and base species)

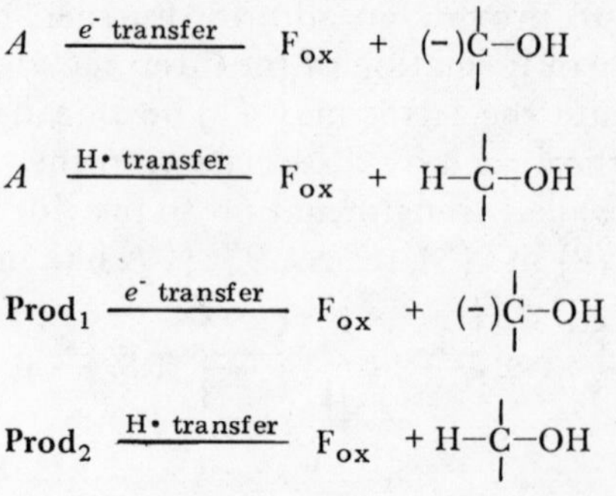

$$A \xrightarrow{\quad e^- \text{ transfer} \quad} F_{ox} + (-)\overset{|}{\underset{|}{C}}-OH$$

$$A \xrightarrow{\quad H \cdot \text{ transfer} \quad} F_{ox} + H-\overset{|}{\underset{|}{C}}-OH$$

$$Prod_1 \xrightarrow{\quad e^- \text{ transfer} \quad} F_{ox} + (-)\overset{|}{\underset{|}{C}}-OH$$

$$Prod_2 \xrightarrow{\quad H \cdot \text{ transfer} \quad} F_{ox} + H-\overset{|}{\underset{|}{C}}-OH$$

Scheme 11

Scheme 11 pertains to carbonyl group reduction by $FH^-$, but a similar scheme would pertain to $FH_2$. In step 1 it is obvious that the free energy content of states **A**, **B**, and **Prod**$_1$ exceed that of **start**. The standard free energies of formation from **start** of **A**, **B**, and **prod**$_1$ are designated as $\Delta F_A^0$, $\Delta F_B^0$ and $\Delta F_{prod_1}^0$, respectively. $\Delta F_A^0 > \Delta F_{prod_1}^0$ by a few or many kilocalories depending on the structre of the carbonyl compound [133]. Also, $\Delta F_A^0$ and $\Delta F_B^0$ may approach each other with $\Delta F_B^0 > \Delta F_A^0$ when the radical species is electronically (inductive and resonance) stabilized as in the case of $-\overset{\cdot}{C}(O^\ominus)CO^-$. The area within the square in Scheme 11 represents a blank board upon which the free energy contour may be drawn for conversion of **start**→**Prod**$_1$ (see references 173-175). The reaction path may be along an edge of the map from **start** → **A** → **Prod**$_1$, or conversely from **start** → **B** → **Prod**$_1$ or directly from **start** → **Prod**$_1$. In the latter case, the position of the transition state may be between **A** and **B** (completely concerted or coupled $e^-$ and $H^+$ transfer) or closer to **A** and **B** (partially coupled). The position of the transition state on the free energy contour map may be determined by the relative $\Delta F^0$ of states **A**, **B**, and **Prod**$_1$. Generally in attempting to predict the position of a transition state on an energy contour map one must not only know the values of $\Delta F^0$ for the four states but also the values of $\Delta F^{\ddagger}$ leading from one state to another. In the present case the values of $\Delta F^0$ should suffice, since $e^-$ and $H^+$ transfer are characterized by nearly identical rate constants of $10^9$ and $10^{10}$ $M^{-1}$ sec$^{-1}$ in the thermodynamically favored driection. Thus if $\Delta F_B^0 \gg \Delta F_A^0 \gg \Delta F_{prod_1}^0$, the reaction path is **start** → **A** → **Prod**$_1$ and so forth. Only in the case where $\Delta F_B^0 \gg \Delta F_{prod_1}^0 \ll \Delta F_A^0$ and the values of $\Delta F_A^0$ and

$\Delta F_B^0$ approach each other is the reaction concerted and does the transition state occur away from the edge of the map. The degree of concertedness depends on the closeness of the values of $\Delta F_A^0$ and $\Delta F_B^0$ and so on. The three possible extreme situations are depicted in the stereoscopic free energy contours of Figure 10. If one considers various combinations of $\Delta F^0$ values and rate-controlling steps and assumes no acid-base catalysis in *step 2*, then it may be predicted that FH$^-$ (and FH$_2$) reduction of $\diagup$C=O could be spontaneous, general-acid, or general-base catalyzed (eq. 96). Many borderline

I  Spontaneous ($v = k_r$[FH$^-$][ $\diagup$C=O])
   (1) $\Delta F_B^0 \gg \Delta F_A^0 \gg \Delta F_{\text{prod}_1}^0$
      Electron transfer in step 1 rate determining. *Step 2* from A or Prod$_1$
   (2) $\Delta F_B^0 \gg \Delta F_A^0 \gg \Delta F_{\text{prod}_1}^0$
      *Step 2* from A is rate determining.

II. General-Acid Catalysis ($v = k_{ga}$[FH$^-$][ $\diagup$C=o][AH])
   (1) $\Delta F_B^0 \gg \Delta F_A^0 \gg \Delta F_{\text{prod}_1}^0$
      H$^+$ transfer in *step 1* rate determining.
   (2) $\Delta F_A^0 \gg \Delta F_B^0 \gg \Delta F_{\text{prod}_1}^0$              (96)
      H$^+$ transfer in *step 1* rate determining.
   (3) $\Delta F_A^0 \gg \Delta F_{\text{prod}_1}^0 \ll \Delta F_B^0; \Delta F_A^0 \cong \Delta F_B^0$
      *Step 1* rate determining.

III. Specific-Acid Catalyzed ($v = k_r$[FH$^-$][ $\diagup$C=O][H$_3$O$^+$])
   (1) $\Delta F_B^0 \gg \Delta F_A^0 \gg \Delta F_{\text{prod}_1}^0$
      *Step 2* from **prod**$_1$ rate determining.
   (2) $\Delta F_A^0 \gg \Delta F_B^0 \gg \Delta F_{\text{prod}_1}^0$
      $e^-$ transfer in *step 1* rate determining.
   (3) $\Delta F_A^0 \gg \Delta F_B^0 \gg \Delta F_{\text{prod}_1}^0$
      *Step 2* from **prod**$_1$ rate determining.
   (4) $\Delta F_A^0 \gg \Delta F_{\text{prod}_1}^0 \gg \Delta F_B^0$
      *Step 1* or *step 2* rate determining.

cases may be expected and an additional complication must be taken into account. Regarding 4a-substitution of F$_{\text{ox}}$ by SO$_3^{2-}$ it has been pointed out that general-acid protonation of N(5) is required becuase of the increased basicity of N(5) in going from F$_{\text{ox}}$ to the 4a-adduct (see discussion on p. 26). The p$K_a$ of ·C—OH species are many orders of magnitude less than the p$K_a$ of H—C—OH [133], thus either 1$e^-$ or H· transfer to ·C—O$^\ominus$ (A) should also require general-acid catalysis. For this reason, case I-2 (eq. 96) is spontaneous but

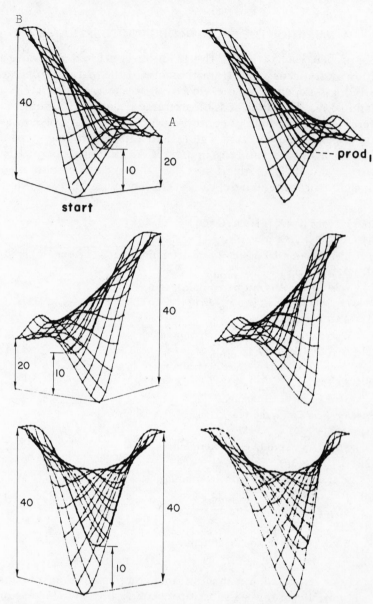

Figure 10. Stereoscopic views of extreme examples of free energy contours for $H^+$ and $1e^-$ transfer processes depicted in *step 1* of Scheme XI. (Top): $\Delta F^0_{\mathbf{A}}$ = 20 kcal $M^{-1}$, $\Delta F^0_{\mathbf{B}}$ = 40 kcal $M^{-1}$, and $\Delta F^0_{\mathbf{Prod_1}}$ = 10 kcal $M^{-1}$. The reaction is not concerted and goes to **prod₁** by way of **A**. (Center): The values of $\Delta F^0$ for states **A** and **B** are reversed; the reaction is not concerted and proceeds from **start** to **prod₁** by way of **B**. (Bottom): $\Delta F^0_{\mathbf{A}}$ = $\Delta F^0_{\mathbf{B}}$ = 40 kcal $M^{-1}$, $\Delta F^0_{\mathbf{prod_1}}$ = 10 kcal $M^{-1}$. The reaction is concerted in $H^+$ and $1e^-$ transfer, and the transtion state is of lower free energy than **A** or **B** and lies midway between the two on the "pass" between **start** and **prod₁**.

appears as specific-acid catalyzed. Also, case III-4 appears as both specific-acid and general-acid catalyzed (i.e., $v = k_r [FH^-][\,\diagdown C=O]$ $[HA][H_3O^+]$). In the model reactions we have investigated to date, either a slightly concerted but mainly specific-acid-catalyzed or spontaneous reduction is in effect. The spontaneous reduction is most likely case I-1.

In the retrograde direction the same type analysis would have to be performed to judge the rate expression that one might obtain. However either $(-)CR_2OH$ or $HCR_2OH$ could act as substrate. At the time of writing, oxidation of alcohol to carbonyl compound has only been well established in those instances when ionization of the H—C bond can occur to provide a resonantly stabilized carbanion, as in the case of $\alpha$-ketols (p. 36). However $CH_3OH$ oxidation by $F_{ox}$ has been established [138], as one would predict from the $\Delta F^0$ for $CH_2O$ + $FH_2 \rightleftharpoons CH_3O + F_{ox}$ (see reaction coordinate diagram of Fig. 7). General-acid catalysis of 4a-addition of $SO_3^{2-}$ has been established (p. 26) and general-acid 4a-addition of $RS^-$ certainly appears as the most likely mechanism (p. 30) in the oxidation of thiols. However it should be clear from the considerations of eq. 96 that if general-acid catalysis should be observed at some date in the oxidation of an alcohol by $F_{ox}$ it would not be sufficient evidence to invoke nucleophilic attack of carbanion to provide a 4a-addition intermediate.

One must presently assume that in enzymatic reactions the catalytic apparatus is capable of generating what would ordinarily be considered high free energy carbanions (eq. 92). In the case of lactate or $\alpha$-amino acid substrate, this would not be highly unlikely since the heterogeneous milieu of the active site could provide charge stabilization and if the carboxyl group remained protonated the carbanion could be resonantly stabilized (see eq. 60). In addition, the free energy of formation of the radical pair $F_{reduced} \cdot + CH_3\overset{..}{C}(OH)$-$COOH$ from $F_{ox} + CH_3CH(OH)COOH$ can be calculated to be rather small ($\Delta F^0 = 6$ kcal $M^{-1}$ at pH 5.0). It should be recalled that carbanions readily undergo $1e^-$ transfer reactions. This is particularly so for carbanions generated from $\alpha$-hydroxycarbonyl compounds (and also nitroalkanes) [110, 111, 176]. In the hypothetical case for which $1e^-$ and H· transfer are competitive, the requirements of microscopic reversibility would be met if the partitioning of the central species $(FH \cdot \cdot \overset{|}{\underset{|}{C}}-OH)$ reflected the percent of its formation from the three initial states (eq. 97). Thus the ratio $k_a : k_b : k_c = k_{-a} : k_{-b} : k_{-c}$. Much exploratory work is required before structure reactivity relationships involving flavins can be discussed in terms of detailed coordinate diagrams. Both extensive kinetic studies and electrochemical

$$FH + {\small \sum} C{=}O \;\rightleftharpoons\; FH\cdots\overset{|}{\underset{|}{C}}OH \quad \overset{k_b}{\underset{k_{-b}}{\rightleftharpoons}} \quad \begin{array}{c} F_{ox} + (-)\overset{|}{\underset{|}{C}}{-}OH \\[1em] \overset{k_c}{\underset{k_{-c}}{\rightleftharpoons}} \quad F_{ox} + H{-}\overset{|}{\underset{|}{C}}{-}OH \end{array} \qquad (97)$$

studies providing the required $E'_o$ values for half-cell reactions are required.

## 5   REACTION OF $FH_2$ WITH $O_2$

Determination of the mechanism of the reaction of $O_2$ with $FH_2$ remains a fascinating and important problem. A useful review, which the author has drawn upon freely, of progress in this area of research has been written by Massey et al. [177]. The reactions in Scheme 12

(a) $FH_2 + O_2 \underset{k_{-1}}{\overset{k_1}{\rightleftharpoons}} FH_2O_2$

(b) $FH_2O_2 \xrightarrow{k_2} FH\cdot + O_2^- + H^+$

(c) $2FH\cdot \underset{k_{-3}}{\overset{k_3}{\rightleftharpoons}} (FH_2 \cdot F_{ox}) \underset{k'_{-3}}{\overset{k'_3}{\rightleftharpoons}} F_{ox} + FH_2$

(e) $FH\cdot + O_2 \xrightarrow{k_4} F_{ox} + O_2^- + H^+$

(f) $FH\cdot + O_2^- + H^+ \xrightarrow{k_5} F + H_2O_2$

(g) $FH_2 + O_2^- + H^+ \xrightarrow{k'_5} FH\cdot + H_2O_2$

(h) $O_2^- + O_2H + H^+ \xrightarrow{k_6} O_2 + H_2O_2$

Scheme 12

have been proposed to account for the kinetics of the oxidation process. In the majority of the studies by Massey and associates, tetraacetylriboflavine has been employed. Estimated rate constants are provided in Table 2. In Scheme 12 the acid-base equilibria for the dihydroflavin and flavin radical, and so on have been ignored and the possible direct breakdown of $FH_2O_2$ to $H_2O_2$ and $F_{ox}$ is not included.

For flavoenzymes the rate of reaction of $FH_2$ with $O_2$ and the eventual products of reaction are markedly controlled by the apoenzyme. Thus the rate of reaction of $O_2$ with a flavoprotein oxidase is as great or greater than that with $FH_2$, while in the case of

### TABLE 2   RATE CONSTANTS ASSOCIATED WITH SCHEME XII

| Rate Constant | Value ($M^{-1}$ sec$^{-1}$) | Reference |
|---|---|---|
| $k_2$ | 10 (sec$^{-1}$) upper limit | 177 |
| $k_3$ | 4 × 10$^8$ (anion radical) | 2 |
|  | 6 × 10$^8$ (neutral radical) |  |
| $k_{-3}$ | 8 × 10$^5$ (anion radical) | 2,4 |
|  | 2 × 10$^6$ (neutral radical) | 178,179 |
| $k_4$ | 1-3 × 10$^8$ (anion radical) | 4 |
|  | <3 × 10$^4$ (neutral radical) |  |
| $k_6$ | 8 × 10$^7$ | 180 |

a flavodehydrogenase it is greatly reduced or in some cases completely repressed. The origin of the slight acceleration in rate for the
oxidase may simply be akin to a solvent effect by the active site.
The decrease in rate for the dehydrogenases strongly suggests steric
hindrance in the region of the 4a- and 5-positions. Note that
5-addition of sulfite (a nucleophile that adds to the 5- and 4a
positions, p. 22) is also sterically forbidden for the dehydrogenases
but not for the oxidases. For both the oxidases and dehydrogenases,
the final products are F$_{ox}$ + H$_2$O$_2$. For the flavoprotein hydroxylases,
on the other hand, one atom of oxygen appears as H$_2$O and the other
is incorporated into substrate. It has been suggested by Massey and
co-workers [177] that the role of the apoprotein is to direct the
mode of decomposition of an FH$_2$O$_2$ adduct. Thus the decomposition of FH$_2$O$_2$ could be directed to F$_{ox}$ plus H$_2$O$_2$ [a reasonable
mechanism for this reaction may be written (eq. 98)] or conceivably
to F· plus O$_2^{-}$.

$$(98)$$

Existing chemical evidence for the intermediacy of the various
species of Scheme 12 are briefly considered. On mixing FH$_2$ and O$_2$
(in the absence or presence of dithionite) a very rapid absorbance
change (400-500 nm) is observed prior to appearance of F$_{ox}$. This

and the following experiment provide evidence for the species $FH_2O_2$ [181]. The rate of reduction of cytochrome c by $FH_2$ is inhibited by $O_2$ in the time period prior to the production of $O_2^{\bar{\cdot}}$, which may be taken as evidence for formation of $FH_2O_2$ or a cytochrome c-$O_2$ complex. The former is favored [177]. The involvement of $F_{ox}$ in the sequence stems both from the known comproportionation reaction (eq. 1) and from the first careful kinetic investigation of the oxidation by Gibson and Hastings ($FMNH_2$, pH 6.3) [182]. These investigators found the reaction to be autocatalytic in production of $F_{ox}$ and suggested the reaction of product $F_{ox}$ with $FH_2$. The autocatalytic involvement of $F_{ox}$ is explained in Scheme 12 by the reversal of reaction c. From the work of Massey and collaborators, it is known that the inclusion of $F_{ox}$ in the initial reaction mixture greatly reduces the lag period. We have found [183] that in the presence of excess (the amount dependent on the particular isoalloxazine examined) oxidized flavin or isoalloxazine, the lag phase may be completely abolished. The detection of $O_2^{\bar{\cdot}}$ is dependent on its readily recognizable epr spectrum. In practice rapid freeze followed by epr spectroscopy has been employed [184]. This technique has been used during the course of the oxidation reaction, and since $O_2^{\bar{\cdot}}$ remains at an appreciable concentration (30% of theory) when all $FH_2$ is converted to $F_{ox}$, it may be employed as a monitor for the disposition of the $O_2^{\bar{\cdot}}$ species. Complete abolition of the lag phase occurs when the reaction of dihydrotetraacetylriboflavin with $O_2$ is followed in the presence of equal concentrations of the $F_{ox}$ species ($5 \times 10^{-5}$ $M$) and in the presence of superoxide dismutase. As anticipated from Scheme 12 superoxide dismutase considerably slows the rate of disappearance of $FH_2$ in either the presence or absence of $F_{ox}$. In the presence of $F_{ox}$ and dismutase, Scheme 12 should reduce to eq. 99. The complexity of the reaction

$$F_{ox} + FH_2 \longrightarrow 2F\cdot$$
$$F\cdot + O_2 \longrightarrow F_{ox} + O_2^- \qquad (99)$$

is great, and reaction of produced $H_2O_2$ with $FH_2$ should probably be taken into consideration. The dependence of the reaction rate on $[O_2]$ has been found to be biphasic in the neutral to alkaline region of pH but first order in $FH_2$ and $O_2$ at acidic pH. The pH maximum for the reaction is around 6.3. Much work remains to be done on the kinetics and mechanism of reaction of reduced flavins and isoalloxazines with $O_2$, particularly in regard to the extremes of pH and the influence of substituent effects. For example, at pH 0 and with XV the reaction is greatly slowed, and it can be shown that dihydroflavin yields flavin radical in a first order process, which then provides F$_{ox}$ again in a first order reaction [183].

The rapidity of the reaction of ground state triplet oxygen with FH$_2$ is of considerable interest. The central ring of 1,5-dihydroiso-alloxazines and the attached N(1)-function constitute at one time a 1,4-dihydropyrazine and an enamine. In the butterfly configuration the 1,4-dihydropyrazine ring will possess N(1)-N(5) orbital-orbital splitting while in the planar conformation it will be anti-aromatic. Thus planar or bent FH$_2$ should be prone to enter into radical mechanisms. This serves to explain the rapidity of the reaction of triplet oxygen with FH$_2$. The author suggests the sequence of events in eq. 100 to account for reactions a, b, and c of Scheme 12. The stable O$_2$

$$FH^{\ominus} + O_2 \rightleftarrows \{FH\cdot\cdot O_2^{\ominus}\} \rightleftarrows FHO_2^- \qquad (100)$$
$$FH\cdot + O_2^{\dot{-}} \qquad F_{ox} + HO_2^-$$

adduct of an enamine (eq. 101) has been isolated in crystalline form [185]. The nature of the postulated FH$_2$O$_2$ adduct has been suggested

$$(101)$$

by Massey and associates to be the 10a- and 4a-peroxides of 1,5-dihydroflavin [**XXXVII** and **XXXVIII**, respectively (eq. 102)]. On the basis

$$(102)$$

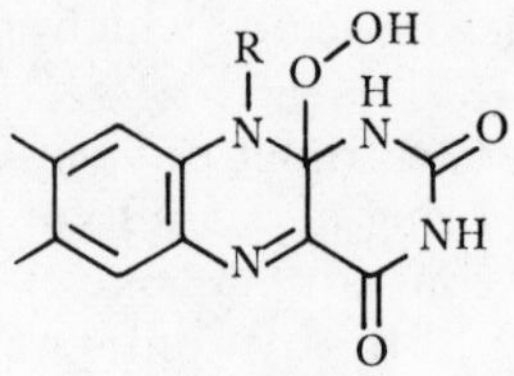

**XXXVII**

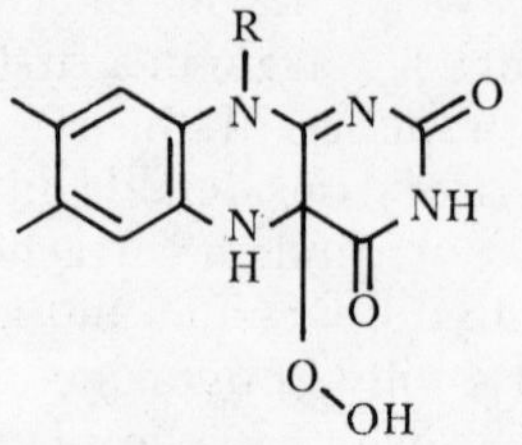

**XXXVIII**

of spectral changes associated with nucleophilic attack of MeO⁻ upon a 1-alkylflavinium compound (**XVA**) [83], Hemmerich and Müller [82] have suggested that 8- and, perferably, 6-peroxy structures be considered for $FH_2O_2$ (**XXXIX** and **XL**, respectively).

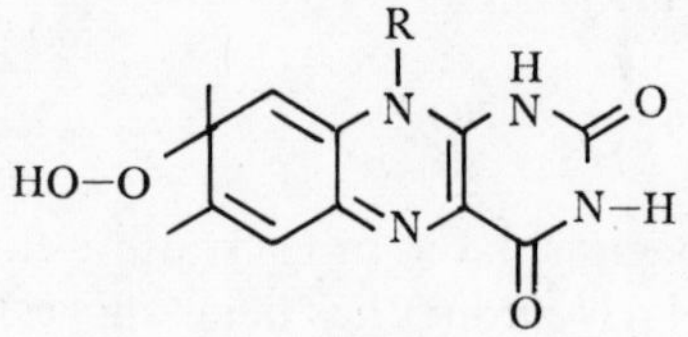

**XXXIX**

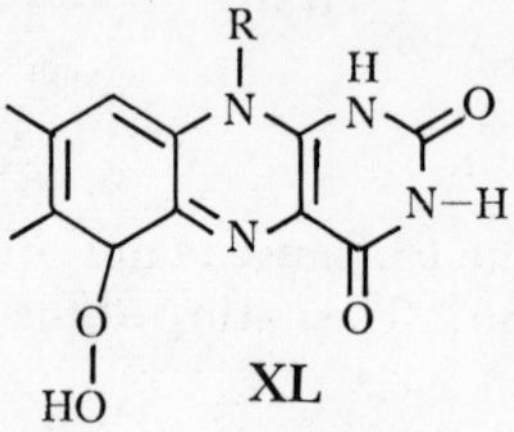

**XL**

It is claimed that triplet molecular $O_2$ reacts with 1,5-dihydroflavin at a rate exceeding that for singlet $O_2$, whereas this is not so for 1-hydro-5-alkylflavins [186]. With the latter compounds singlet $O_2$ is reported to provide an adduct that can be trapped by $N_3^-$, isolated, and subjected to mass spectral analysis. The structure of **XLI** was assigned to the trapped adduct,

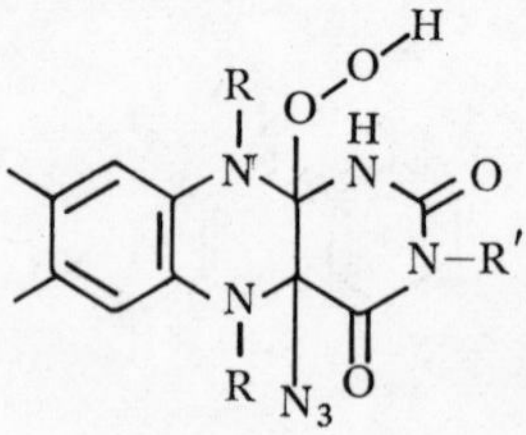

**XLI**

although the means of determining the position of attachment of the $O_2$ moiety (4a versus 10a)—as well as other features of this study—are not clear. Certainly the steric availability of the 10a-position is not essential to the reaction of $FH_2$ with $O_2$. For the isoalloxazine **IX**, approach to the 10a-position is hindered by the 2′- and 6′-methyl substituents, and at pH 6.4 in the presence of $\sim 10^{-7}$ $M$ superoxide dismutase and equimolar ratios of oxidized and reduced flavin (5 $\times$ 10$^{-5}$ $M$), we find [183] the stopped-flow rate constants for O$_2$ oxidation of 1,5-dihydrolumiflavin and dihydro-**IX** to be comparable. These results establish that the rate of formation of **XLII** is as great as that for the formation of **XLIII** and that a 10a-peroxy adduct is not involved, or at least not essential. Orf and Dolphin

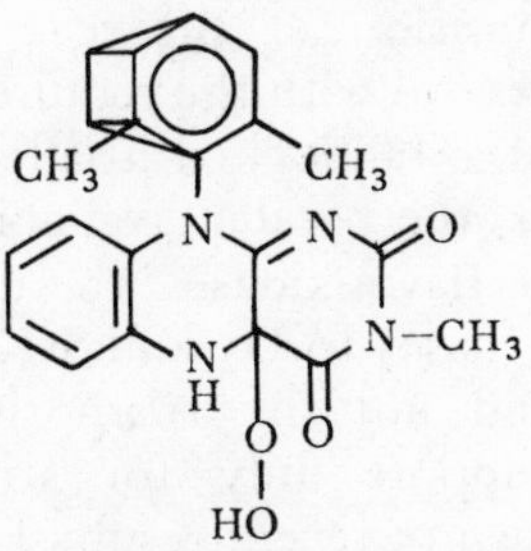

XLII 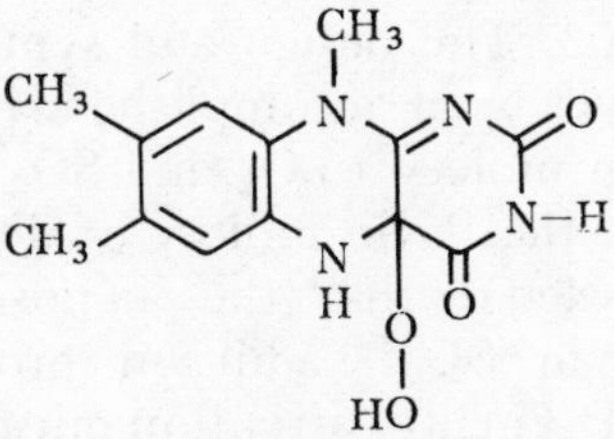          XLIII

propose [51] that a peroxide dihydroflavin adduct is formed by way of a dioxetane intermediate (eq. 103). They note the analogy between the

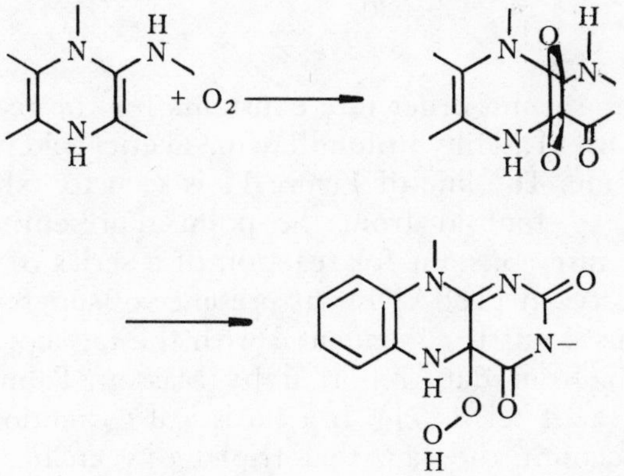

(103)

strongly electron-donating tetraaminoethylenes [187] and FH$_2$ and the fact that octamethyl tetraaminoethylene reacts with O$_2$ in the presence of a proton donor to provide a urea. The best evnisioned mechanism for this reaction is considered to proceed by way of a dioxetane [188] (eq. 104). The N(1)-proton of

$$(104)$$

the postulated dihydroflavo-dioxetane (eq. 103) is suggested to provide an alternate means of oxetane ring opening. The unimportance of the availability of the 10a-position in the reaction of dihydroisoalloxazine with $O_2$ is not in accord with the requirements of eq. 103. The design and synthesis of a 4a-blocked isoalloxazine has not yet been accomplished. However, the parallel availability of the flavin moiety to $O_2$ and $SO_2^{2-}$ in the flavooxidases, the unavailability of the flavin moiety of dehydrogenases to $O_2$ and $SO_3^{2-}$, and the knowledge that the 4a-position and not the 10a-position is involved in $SO_3^{2-}$ addition provide support only for structure **XXXVIII**. Finally, attention must be drawn to recent results [188A] which point out the similarity in the reaction of ($FH_2$ + $FH^-$) with triplet $O_2$ and nitroxides (eq. 104A). In Figure 11 there is plotted

$$(104A)$$

the log of the second order rate constants for the reaction of three nitroxides with 1,5-dihydrolumiflavin-3-acetic acid versus the $E_{1/2}$ of the nitroxide. The line of Figure 11 is seen to exhibit a negative deviation of *ca* tenfold from the point representing the average second order rate constant for reaction of a series of dihydroflavins with triplet oxygen [188A] in the presence of superoxide dismutase and $F_{ox}$. (This point is coincidental with the rate constant that may be calculated from data reported by Massey, Palmer and Ballou [Figs. 7 and 8 of ref. 177]). If a statistical correction is made that takes into account the fact that triplet oxygen has two unshared electrons and the nitroxides have but one, the oxygen rate is but *ca* 5-fold greater than anticipated from its $E_{1/2}$. This small factor is easily accounted for by the steric requirements of the nitroxides as compared to oxygen. Further, when empolying a series of 1,5-dihydroflavins and a single nitroxide the log of the second order

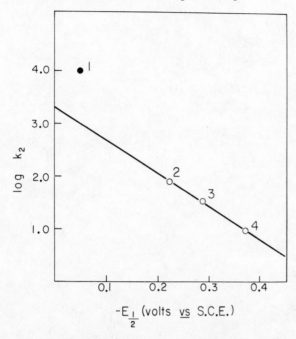

Figure 11.  A plot of the log of the second order rate constants ($M$ sec$^{-1}$) for reaction of oxygen (1) with a series of 1,5-dihydroflavins and the nitroxides 2,2,6,6-tetramethyl-4-piperidinol-1-oxy (2), 3-carbamoyl-2,2,5,5-tetramethylpi-peridinyl-1-oxy (3), 3-carboxy-2,2,5,5-tetramethylpiperidinyl-1-oxy (4) with 1,5-dihydrolumiflavin-3-acetic acid [123 A].

rate constant is found to be a linear function of the $E_{\frac{1}{2}}$ of the dihydroflavin [188A]. The rate constants for reaction of the iso-alloxazine IX fits this line. Clearly 9a- or 10a-positions cannot be positions for the reaction of the bulky nitroxides with dihydroflavin. Reaction of nitroxides with FMeH and FMe$^-$ yield FMe$\cdot$ plus hydro-xylamine. These results establish that the rate of $1e^-$ transfer from 1,5-dihydroflavin to oxygen is in accord with the latter's reduction potential.

It is most likely that 8- and 6-peroxy structures (**XXXIX** and **XL**) are not involved in the reoxidation of dihydroisoalloxazine in aqueous solution. Referring to the reactions of eqs. 36 and 37 it is seen that the reaction of HO$_2^-$ with **XVII** to provide the 8-hydro-xyisoalloxazine (**XVIII**) should pass through the same intermediate that would be obtained on addition of O$_2$ to the 8-position of reduced **XVII** (eq. 105). In experiments carried out at pH 5.0 and

(105)

9.5 [91], 20 cycles of alternate dithionite reduction and $O_2$ oxidation of reduced **XLIV** gave a 94% yield of unaltered **XVII**. No absorbance at 480 nm (**XVIII**) could be detected at the end of either experiment. Therefore the dihydro form of **XVII** (i.e., **XVIIA**) does not undergo oxidation by addition of $O_2$ at either the 6- or 8-position.

At present there is much interest in the observation that several external flavoprotein hydroxylases evidence spectrally identifiable intermediates when $O_2$ is allowed to react with the reduced enzyme substrate complexes. Such intermediates have been seen in the case of $p$-hydroxybenzoate hydroxylase ($\lambda_{max}$ 385 nm, $\epsilon = 8500$ $M^{-1}$ cm$^{-1}$ [189]), melilotate hydroxylase (spectra similar to that formed with the $p$-hydroxybenzoate enzyme [151]), and phenol hydroxylase ($\lambda_{max}$ 380 nm [190]). It has been suggested that the observed intermediates are covalent adducts of $O_2$ and reduced flavin (structure **XXXVIII** is favored, but the spectral characteristics are not totally unambiguous). It has been shown in an elegant experiment [191] that mixing bacterial luciferase, FMNH$_2$, and $O_2$

in 50% ethylene glycol-phosphate buffer for 10 sec at 4°C followed by quenching at -20°C and removal of unbound FMN by chromatography at -20°C provides a substance possessing a $\lambda_{max}$ at 355 nm. Maximum absorptions in this range are provided by 4a-adducts (see, for example, **XX**). Warming this species provides $F_{ox}$ + $H_2O_2$.

Compelling evidence for the formation of a 4a-peroxy adduct on oxidation of dihydroflavin has been obtained by Mr. Kemal [192] in the author's laboratory (eq 106). When KI is added to a fresh solution of **XLIV** in methanol, $KI_3^-$ immediately appears in 85 to 90%

$$(106)$$

**XLIV**

yellow crystals
$\lambda_{max}$ (methanol)

370 nm ($\epsilon$ 8500 M$^{-1}$ cm$^{-1}$)
308 nm ($\epsilon$ 6600 M$^{-1}$ cm$^{-1}$)

of theory. Reaction of 1,5-dihydro-3,5-dimethyllumiflavin with O$_2$ in methanolic solution provides a product which reacts on mixing with KI to yield $KI_3^-$ in > 70% yield based on dihydroflavin employed. It would appear, therefore, that **XLIV** can be generated by reaction of $HO_2^-$ with the oxidized N(5)-methylflavin or by reaction of O$_2$ with reduced N(5)-methyl flavin.

An investigation of the ability of **XLIV** to insert oxygen into electron rich aromatic nuclei is in progress.

On addition of hexanal to a dioxane solution of **XLIV** light emission immediately ensues. In the presence of excess H$_2$O$_2$ and RCHO (formaldehyde, hexanal, heptanal, decanal, etc.) over **XLIV**, the light emission continues for hours and is visible to the dark adapted eye. The tentative partial mechanism of 107 is suggested. The fluoroescence emission spectra of **XLV** (excited at 354 nm) is characterized by a $\lambda_{max}$ at 460 nm (frozen H$_2$O-ethylene glycol) which is close to that ($\sim$ 490 nm) reported for bacterial luciferase

$$(107)$$

[193]. In aqueous solution, the decrease in light emission has been found to parallel the decrease in concentration of **XLIV**. Thus, the reaction of **XLIV** with aldehydes may serve as a useful model for bacterial luciferase.

## 6 EPILOGUE

The author was once told that there are two times one should write a review chapter. When one enters a field and when one leaves. In the first instance, one can point out what is not known and be excused for imaginative speculation. In the latter instance, one may summarize what is well established. Physical organic studies of flavin catalysis have proceeded beyond the beginnings but much still remains to be done. The author has, in instances, freely speculated beyond what is presently known. For any foresight he takes complete credit. For any fallacious arguments he assumes no blame. Since chemistry is an experimental science, worthwhile advances do not often come about solely by the contemplation of one's navel.

## ACKNOWLEDGMENT

This work was supported by a grant from the National Science Foundation.

# REFERENCES

1. F. Müller, P. Hemmerich, and A. Ehrenberg, in *Flavins and Flavoproteins*, H. Kamin, Ed., University Park Press, Baltimore, 1971, p. 107.
2. E. J. Land and A. J. Swallow, *Biochemistry*, 8, 2117 (1969).
3. P. Hemmerich, C. Veeger, H. C. S. Wood, *Angew. Chem. (Int. Ed. Engl.)*, 4, 671 (1965).
4. A. Ehrenberg, F. Müller and P. Hemmerich, *Eur. J. Biochem.*, 2, 286 (1967).
5. R. D. Draper and L. L. Ingraham, *Arch. Biochem. Biophys.*, 125, 802 (1968).
6. O. Gawron, A. Rampol and P. Johnson, *J. Am. Chem. Soc.*, 94, 5396 (1972).
7. H. Beinert, *J. Am. Chem. Soc.*, 78, 5323 (1956).
8. F. Müller, P. Hemmerich, A. Ehrenberg, G. Palmer, and V. Massey, *Eur. J. Biochem.*, 14, 185 (1970).
8a. A. Ehrenberg, F. Müller, and P. Hemmerich, *Eur. J. Biochem.*, 2, 286 (1967).
8b. Q. H. Gibson, V. Massey, and N. M. Atherton, *Biochem. J.*, 85, 369 (1962).
8c. V. Massey and G. Palmer, *Biochemistry*, 5, 3181 (1966).
9. J. H. Swinehart, *J. Am. Chem. Soc.*, 87, 904 (1965).
10. B. G. Barman and G. Tollin, *Biochemistry*, 11, 4670 (1972).
10a. T. C. Bruice, unpublished results; presented in simpler form by P. Hemmerich, *Helv. Chim. Acta*, 47, 464 (1964); *see also ref.* 8a.
11. H. Beinert, *Enzymes*, 2nd ed. 2, Chap. 10 (1960); Academic Press, New York, 1960, Chap. 10.
12. T. C. Bruice, L. Main, S. Smith, and P. Y. Bruice, *J. Am. Chem. Soc.*, 93, 7327 (1971).
13. L. P. Hammett, in *Physical Organic Chemistry*, McGraw-Hill, New York, 1940, p. 188.
14. M. A. Slifkin, in *Charge Transfer Interactions of Biomolecules*, Academic press, New York, 1971, Chap. 7.
15. V. Massey and S. Ghisla, *Ann. N. Y. Acad. Sci.*, 227, 446 (1974).
16. F. Müller and V. Massey, *J. Biol. Chem.*, 244, 4007 (1969).
17. L. Tauscher, S. Ghisla, and P. Hemmerich, *Helv. Chim. Acta*, 56, 630 (1973).
17a. S. Ghisla, V. Massey, J. M. Lhoste, and S. G. Mayhew, *Biochemistry*, 13, 589 (1974).
17b. G. Blankenhorn, private communication.
18. M. Brüstlein and T. C. Bruice, unpublished.
19. L. Main, G. J. Kasperek, and T. C. Bruice, *J. Chem. Soc.*, 1972, 847; *ibid.*, Biochemistry, 11, 3991 (1972).
20. R. F. Williams and T. C. Bruice, unpublished.
21. R. M. Burnett, G. D. Darling, D. S. Kendal, M. E. LeQuesne, S. G. Mayhew, W. W. Smith, and M. L. Ludwig, *J. Biol. Chem.*, 249, 4383 (1974).

22. R. D. Draper and L. L. Ingraham, *Arch. Biochem. Biophys.*, 139, 265 (1970).

23. J. M. Gillard and G. Tollin, *Biochem. Biophys. Res. Commun.*, 58, 328 (1974).

24. T. P. Singer and D. E. Edmondson, *FEBS Lett.*, 42, 1 (1974).

25. W. H. Walker, T. P. Singer, *J. Biol. Chem.*, 245, 4224 (1970).

26. W. H. Walker, T. P. Singer, S. Ghisla, and P. Hemmerich, *Eur. J. Biochem.*, 26, 279 (1972).

27. H. Möhler, M. Brühmüller, and K. Decker, *Eur. J. Biochem.*, 29, 152 (1972).

28. D. R. Patex and W. R. Frisell, *Arch. Biochem. Biophys.*, 150, 347 (1972).

29. W. C. Kennery, D. E. Edmondson, and T. P. Singer, *Biochem. Biophys. Res. Commun.*, 57, 106 (1974).

30. W. C. Kenney, D. E. Edmondson, T. P. Singer, D. J. Steenkamp, and J. C. Schabort, *FEBS Lett.*, 41, 111 (1974).

31. T. P. Singer and W. C. Kenny, in *Vitamins and Hormones*, in press, 1974.

32. W. C. Kenney, D. Edmondson, R. Seng, and T. P. Singer, *Biochem. Biophys. Res. Commun.*, 52, 434 (1973).

33. D. E. Edmondson, *Biochemistry*, 13, 2817 (1974).

34. S. Ghisla and S. G. Mayhew, *J. Biol. Chem.*, 248, 6568 (1973).

35. C. D. Whitfield and S. G. Mayhew, *J. Biol. Chem.*, 249, 2801, 2811 (1974).

36. S. G. Mayhew, C. D. Whitfield, S. Ghisla, and M. S. Jörns, *Eur. J. Biochem.*, 44, 579 (1974).

37. G. Schöllnhammer and P. Hemmerich, *Z. Naturforsch.*, 27b, 1030 (1972).

38. G. Schöllnhammer and P. Hemmerich, *Eur. J. Biochem.*, 44, 561 (1974).

39. R. Miura, K. Matsui, K. Hirotsu, A. Shimada, M. Takatsu, and S. Otani, *Chem. Commun.*, 1973, 703.

40. P. Hemmerich, G. Nagelschneider, *FEBS Lett.*, 8, 69 (1970).

41. M. S. Flashner and V. Massey in *Molecular Mechanisms of Oxygen Activation*, O. Hayaishi, Ed., Academic Press, New York, 1974, Chap. 7.

41a. O. Lockridge, V. Massey, and P. A. Sullivan, *J. Biol., Chem.*, 247, 8097 (1972).

42. D. K. Rohrbaugh and G. A. Hamilton, private communication.

43. H. I. X. Mager and W. Berends, *Tetrahedron*, 30, 917 (1974) and references therein.

44. S. A. Goscin and I. Fridovich, *Arch. Biochem. Biophys.*, 153, 778 (1972).

45. F. Haber and J. Weiss, *Proc. R. Soc. Lond.*, A147, 332 (1934).

46. L. M. Dorfman, I. A. Taub, and R. F. Bühler, *J. Chem. Phys.*, 36, 3051 (1962).

47. J. M. McCord and I. J. Fridovich, *J. Biol. Chem.*, 244, 6049 (1969).

48. S. Strickland and V. Massey in *Oxidases and Related Redox Systems*, T. E. King, H. S. Mason, and M. Morrison, Eds., University Park Press, Baltimore.

49. J. A. Peterson, *Arch. Biochem. Biophys.*, 144, 678 (1971).

50. G. A. Hamilton, in *Progress in Bioorganic Chemistry*, Vol. I, E. T. Kaiser and F. J. Kezdy, Eds., Wiley-Interscience, New York, 1971, p. 83.

51. H. W. Orf and D. Dolphin, *Proc. Nat. Acad. Sci. (U. S.)*, 71, 2646 (1974).

52. D. M. Jerina, D. R. Boyd and J. W. Daly, *Tetrahedron Lett.*, 457 (1970).

53. C. Kanedo, I. Yokoe, S. Yamada and M. Ishivawa, *Chem. Pharm. Bull. (Tokyo)*, 14, 1316 (1966).

54. A. Alkaitis and M. Calvin, *Chem. Commun.*, 29 (1968).

55. A. F. Hegarty and T. C. Bruice, *J. Am. Chem. Soc.*, 91, 4924 (1969).

56. A. F. Hegarty and T. C. Bruice, *J. Am. Chem. Soc.*, 92, 6575 (1970).

57. B. L. VanDauren, I. Bekersly, and M. Lefar, *J. Org. Chem.*, 29, 686 (1964).

58. E. Boyland and P. Sims, *Biochem. J.*, 90, 391 (1964).

59. D. M. Jerina, J. W. Daly, and B. Whitkop, *Biochemistry*, 10, 366 (1971).

60. S. Kaufman and D. B. Fisher, in *Molecular Mechanisms of Oxygen Activation*, O. Hayaishi, Ed., Academic Press, New York, 1974, p. 285.

61. J. A. Blair and A. J. Pearson, *J. Chem. Soc.* (Perkins II) 1974, 80.

62. G. Garoff, J. W. Daly, D. M. Jerina, J. Renson, B. Witkop, and S. Udenfriend, *Science*, 157, 1524 (1967).

63. D. M. Jerina, J. W. Daly, B. Witkop, P. Zaltzman-Nirenberg, and S. Udenfriend, *Arch. Biochem. Biophys.*, 128, 176 (1968).

64. E. Boyland and P. Sims, *Biochem. J.*, 96, 788 (1965).

65. D. M. Jerina, J. W. Daly, and B. Witkop, Experientia, 28, 1129 (1972).

66. D. M. Jerina, J. W. Daly, and B. Witkop, P. Zaltzman-Nirenberg, and S. Udenfriend, *J. Am. Chem. Soc.*, 90, 6525 (1968).

67. G. J. Kasperek and T. C. Bruice, *J. Am. Chem. Soc.*, 94, 198 (1972).

68. G. J. Kasperek, T. C. Bruice, H. Yagi, and D. M. Jerina, *Chem. Commun.*, 784 (1972).

69. G. J. Kasperek, T. C. Bruice, H. Yagi, N. Kaubisch, and D. M. Jerina, *J. Am. Chem. Soc.*, 94, 7876 (1972).

70. G. J. Kasperek, P. Y. Bruice, T. C. Bruice, H. Yagi, and D. M. Jerina, *J. Am. Chem. Soc.*, 95, 6041 (1973).

70a. J. W. Hastings, A. Eberhard, T. O. Baldwin, M. Z. Nicoli, T. W. Cline, and K. H. Nealson in *Chemiluminescence and Bioluminescence*, M. J. Cormier, D. M. Hercules, and J. Lee, Eds., Plenum Press, New York City, 1973, p. 375.

70b. M. DeLuca and M. E. Dempsey, *Biochem. Biophys. Res. Commun.*, 46, 117 (1970); M. DeLuca and M. E. Dempsey in *Chemiluminescence and Bioluminesecne*, M. J. Cormier, D. M. Hercules, and J. Lee, Eds., Plenum Press, New York City, 1973, p. 345.

70c. T. A. Hopkins, H. H. Seliger, E. H. White, and M. W. Cass, *J. Am. Chem. Soc.*, 89, 9148 (1967); E. H. White, T. A. Hopkins, H. H. Seliger, and E. Rapaport, *J. Am. Chem. Soc.*, 91, 2178 (1969).

71. F. H. Westheimer, in *The Mechanisms of Enzyme Action*, W. D. McElroy and B. Glass, Eds., The Johns Hopkins Press, Baltimore, 1954, p. 32.

72. D. S. Coffey and L. Hellerman, *Biochemistry*, 3, 394 (1964).

73. P. Hemmerich and M. Shuman Jorns, in *Enzymes Structure and Function*, C. Veeger, J. Drenth, and R. A. Oasterboan, Eds., North Holland, Amsterdam, 1973, p. 95.

74. S. B. Smith and T. C. Bruice, *J. Am. Chem. Soc.*, 97, 2875, (1974).

75. R. H. DeWolfe and R. C. Newcomb, *J. Org. Chem.*, 36, 3870 (1971).

76. D. A. Wadke and D. E. Guttman, *J. Pharm. Sci.*, 55, 1088, 1363 (1966).
77. D. E. Guttman and T. E. Platek, *J. Pharm. Sci.*, 56, 1432 (1967).
78. Y. Yokoe and T. C. Bruice, *J. Am. Chem. Soc.*, 97, 450 (1975).
79. K. H. Dudley and P. Hemmerich, *J. Org. Chem.*, 32, 3049 (1967).
80. S. Ghisla, U. Hartmann, P. Hemmerich, and F. Müller, *Ann. Chem.*, 1973, 1388.
81. S. Ghisla, private communication.
82. P. Hemmerich and F. Müller, *Ann. N.Y. Acad. Sci.*, 212, 13 (1973).
83. F. Müller, in *Flavins and Flavoproteins*, H. Kamin, Ed., University Park Press, Baltimore, 1971, p. 363.
83a. F. Müller and P. Hemmerich, private communication.
84. F. Müller, *Z. Naturforsch.*, 27b, 1023 (1972).
85. V. Massey, F. Müller, R. Feldberg, M. Schuman, P. A. Sullivan, L. G. Howell, S. G. Mayhew, and R. H. Mathews, *J. Biol. Chem.*, 244, 3999 (1969).
85. L. Hevesi and T. C. Bruice, *Biochemistry*, 12, 290 (1973).
87. P. Gassman and G. A. Campbell, *J. Am. Chem. Soc.*, 94, 3891 (1972).
88. F. J. Bullock and O. Jardetsky, *J. Org. Chem.*, 30, 2066 (1965).
89. K. Bowden and R. Stewart, *Tetrahedron*, 21, 261 (1965).
90. F. M. Vainstein, I. I. Kukhtendo, E. I. Tomilenko, and E. A. Shilov, *Zh. Org. Khim.*, 3, 1654 (1967).
91. S. B. Smith, M. Brüstlein, and T. C. Bruice, *J. Am. Chem. Soc.*, 96, 3696 (1974).
92. T. C. Bruice, L. Hevesi, and S. Shinkai, *Biochemistry*, 12, 2038 (1973).
93. This type situation has been discussed under the general heading of the Libido Rule by W. P. Jencks, *Chem. Rev.*, 72, 705 (1972).
94. C. H. Williams, B. D. Burleigh, Jr., S. Ronchi, L. D. Arscott, and E. T. Jones, in *Flavins and Flavoproteins*, H. Kamin, Ed., University Park Press, Baltimore, 1971, p. 295.
95. V. Massey, Q. H. Gibson, and C. Veeger, *Biochem. J.*, 77, 341 (1960).
96. V. Massey and C. Veeger, *Biochem. Biophys. Acta*, 48 33 (1961).
97. V. Massey and C. H. Williams, Jr., *J. Biol. Chem.*, 240, 4470 (1965).
98. G. Zanetti and C. H. Williams, Jr., *J. Biol. Chem.*, 242, 5232 (1967).
99. L. Thelander, *Eur. J. Biochem.*, 4, 407 (1968).
100. S. Rondri and C. H. Williams, Jr., *J. Biol. Chem.*, 242, 2083 (1972).
101. B. D. Burleigh, Jr., and C. H. Williams, Jr., *J. Biol. Chem.*, 247, 2077 (1972).
102. V. Massey, R. G. Mathews, G. P. Foust, L. G. Howell, C. H. Williams, G. Zanetti, and S. Ronchi, in *Pyridine Nucleotide-Dependent Dehydrogenases*, H. Sund, Ed., Springer-Verlag, Berlin, p. 393.
102a. D. R. Sanadi, *J. Biol. Chem.*, 236, 2317 (1961).
103. I. M. Gascoigne and G. K. Radda, *Biochem. Biophys. Acta*, 131, 498 (1967).
104. E. Loechler and T. Hollocher, private communication.
105. T. C. Bruice, S. J. Benkovic, *Bioorganic Mechanisms*, Vol. I, Benjamin, New York, 1966, Chap. 1.

106. T. C. Bruice, in *Enzymes*, 3rd ed. 2, 217 (1970).
107. D. O. Carr and D. E. Metzler, *Biochim. Biophys. Acta*, 205, 63 (1970).
108. T. C. Bruice, S. Shinkai, and J. Taulane, unpublished.
109. S. Shinkai, T. Kunitake and T. C. Bruice, *J. Am. Chem. Soc.*, 96, 7140 (1974).
110. N. Kornblum, *Trans. N.Y. Acad. Sci.*, 29, 1 (1966).
111. R. C. Kerber, G. W. Urry, and N. Kornblum, *J. Am. Chem. Soc.*, 87, 4520 (1965).
112. L. E. Brown and G. A. Hamilton, *J. Am. Chem. Soc.*, 92, 7225 (1970).
113. D. J. T. Porter, J. G. Voet, and H. J. Bright, *J. Biol. Chem.*, 247, 1951 (1972); *J. Biol. Chem.*, 248, 4400 (1973).
114. D. Clerin and T. C. Bruice, *J. Am. Chem. Soc.*, 96, 5571 (1974).
115. N. Kornblum, R. C. Kerber, and G. W. Urry, *J. Am. Chem. Soc.*, 86, 3904 (1964).
116. G. A. Russell and W. C. Dannen, *J. Am. Chem. Soc.*, 88, 5663 (1966).
117. N. Kornblum, R. E. Michel, and R. C. Kerber, *J. Am. Chem. Soc.*, 88, 5660, 5662 (1966).
117a. P. Hemmerich, private communication.
118. W. R. Knappe, Ph.D. Dissertation, University of Konstanz, Germany, 1971.
119. W. Haas and P. Hemmerich, *Z. Naturforsch*, 27b, 1035 (1972).
120. W. H. Walker, P. Hemmerich, and V. Massey, *Eur. J. Biochem.*, 13, 258 (1970).
121. G. Blankenhorn, S. Ghisla, and P. Hemmerich, *Z. Naturforsch.*, 27b, 1038 (1972).
122. S. Shinkai and T. C. Bruice, *J. Am. Chem. Soc.*, 95, 7526 (1973).
123. R. F. Williams, S. Shinkai, and T. C. Bruice, *Proc. Nat. Acad. Sci. (US)*, 72, 1763 (1975); *J. Am. Chem. Soc.*, (1976).
124. K. Dimroth and J. von Laufenberg, *Chem. Ber.*, 105, 1044 (1972).
125. W. Ried and W. Käppler, *Ann. Chem.*, 687, 183 (1965).
126. J. Ciabattoni and G. A. Berchtold, *J. Org. Chem.*, 31, 1336 (1966).
127. J. N. S. Tam, G. N. Vaughan, M. P. Mertes, G. S. Rork, and I. H. Pitman, unpublished results.
128. D. V. Santi and C. F. Brewer, *J. Am. Chem. Soc.*, 90, 6236 (1968).
129. D. V. Santi, C. F. Brewer, and D. G. Farber, *J. Heterocycl. Chem.*, 1, 903 (1970).
130. W. P. Jencks, *Catalysis in Chemistry and Enzymology*, McGraw-Hill, 1969, Chapter 10.
131. B. M. Dunn and T. C. Bruice, *Adv. Enzymol.*, 37, 1 (1973).
132. Y. Yano and T. C. Bruice, *J. Amer. Chem. Soc.*, 97, 5263 (1975)
132a. A. Streitwieser, in *Molecular Orbital Theory for Organic Chemistry*, Wiley, New York, 1961, p. 273.
132b. N. Trinajstic, *J. Mol. Struct.*, 8, 236 (1971).
132c. J. W. Lown, M. H. Akhtar, and R. S. McDaniel, *J. Org. Chem.*, 39, 1998 (1974).
132d. R. Hoffmann, *Acc. Chem. Res.*, 4, 1 (1971).

132e. G. A. Razuvaev, G. A. Abakumov, and V. A. Pestunovich, *J. Struct. Chem.*, 5, 274 (1964).

133. P. S. Rao and E. Hayon, *J. Am. Chem. Soc.*, 96, 1287 (1974).

134. J. Lilie, G. Beck, and A. Henglein, *Ber. Bunsenges. Phys. Chem.*, 75, 458 (1971).

135. M. L. Poutsma, in *Free Radicals*, Vol. II, J. K. Kochi, Ed., Wiley-Interscience, New York, 1973, p. 122.

136. K. Adamic, D. F. Bowman, and K. U. Ingold, *J. Oil Chem. Soc.*, 47, 109 (1970).

137. G. C. Levy and J. D. Cargioli, *Tetrahedron Let.*, 12, 919 (1970).

138. T. C. Bruice, R. F. Williams, and S. Shinkai, *Proc. Nat. Acad. Sci. U.S.*, 72, 1763 (1975).

139. P. S. Rao and E. Hayon, *J. Phys. Chem.*, 77, 2753 (1973).

140. M. J. Gibian and J. A. Rynd, *Biochem. Biophys. Res. Commun.*, 34, 594 (1969).

141. G. R. Drysdale, M. J. Spiefel and P. Strittmatter, *J. Biol. Chem.*, 236, 2333 (1961).

142. P. Strittmatter, in *Enzymes*, (2nd ed.) 8, chap. 5 (1963).

143. E. M. Kosower, in *Flavins and Flavoproteins*, E. Slater, Ed., Elsevier, Amsterdam, 1966, p. 1.

144. M. Brüstlein and T. C. Bruice, *J. Am. Chem. Soc.*, 94, 6548 (1972).

145. J. J. Steffens and D. M. Chipman, *J. Am. Chem. Soc.*, 93, 6694 (1971).

146. D. J. Creighton, J. Hajdu, G. Mooser, and D. S. Sigman, *J. Am. Chem. Soc.*, 95, 6855 (1973).

147. A. Ohno and N. Kito, *Chem. Lett.*, *(Japan)*, 1972, 369.

148. N. Kito, Y. Ohmshi, and A. Ohno, *Proc. Conference Organic Radicals*, *14th Fukuoka*, 1973, p. 16.

149. D. J. Porter, G. Blankenhorn, and L. L. Ingraham, *Biochem. Biophys. Res. Commun.*, 52, 447 (1973).

150. T. Uedo and M. J. Coon, *J. Biol. Chem.*, 247, 5010 (1972).

151. S. Strickland and V. Massey, *J. Biol. Chem.*, 248, 2953 (1973).

152. G. Blankenhorn, *Eur. J. Biochem.*, 50, 351 (1975).

153. G. Blankenhorn, Biochemistry 14, 3172 (1975).

154. O. Sinanoglu, *Ann. N.Y. Acad. Sci.*, 158, 308 (1969).

155. J. L. Cohen and K. A. Connors, *J. Am. Chem. Soc.*, 91, 3597 (1969).

156. K. A. Connors and S. Sun, *J. Am. Chem. Soc.*, 93, 7239 (1971).

157. D. E. Edmondson, B. Barman, and G. Tollin, *Biochemistry*, 11, 1133 (1972).

158. M. Sun and P. S. Song, *Biochemistry*, 12, 4663 (1973).

159. J. Fisher and C. Walsh, *J. Am. Chem. Soc.*, 96, 4345 (1974).

160. M. S. Jorns and L. B. Hersh, *J. Am. Chem. Soc.*, 96, 4012 (1974).

161. S. Grossman, D. E. Edmondson, E. B. Kerney, and T. P. Singer, private communication.

161a. B. A. Averill, A. Schonbrunn, R. H. Abeles, L. T. Weinstock, C. C. Cheng, J. Fisher, R. Spencer, and C. Walsh, *J. Biol. Chem.*, 250, 1603 (1975).

161b. J. Fisher, R. Spencer, and C. Walsh, to be published.

162. R. E. Cathou and J. M. Buchanan, *J. Biol. Chem.*, **238**, 1746 1963).

163. R. L. Kisliak, *J. Biol. Chem.*, **238**, 397 (1963).

164. V. S. Gupta and F. M. Huennekins, *Arch. Biochem. Biophys.*, **120**, 712 (1967).

165. S. J. Benkovic and W. P. Bullard, in *Progress in Bioorganic Chemistry*, Vol. 2, E. T. Kaiser and F. J. Kézdy, Eds., Wiley, New York, 1973, p. 133.

166. D. Mauzerall and F. H. Westheimer, *J. Am. Chem. Soc.*, 77, 2261 (1955).

167. S. J. Benkovic, W. P. Bullard, and P. A. Benkovic, *J. Am. Chem. Soc.*, 94, 7542 (1972).

168. J. A. Rynd and M. J. Gibian, *Biochem. Biophys. Res. Commun.*, 41, 1097 (1970).

169. C. T. Walsh, A. Shonbrunn, and R. H. Abeles, *J. Biol. Chem.*, 246, 6855 (1971).

170. C. T. Walsh, E. Krodel, V. Massey, and R. H. Abeles, *J. Biol. Chem.*, 248, 1946 (1973).

171. C. T. Walsh, O. Lockridge, V. Massey, and R. H. Abeles, *J. Biol. Chem.*, 248, 7049 (1973).

172. H. Bright, private communication.

173. R. A. More O'Ferrall, *J. Chem. Soc. B*, 274 (1970).

174. W. P. Jencks, *Chem. Rev.*, 72, 705 (1972).

175. B. M. Dunn, Int. *J. Chem. Kinet.*, 6, 143 (1974).

176. G. A. Russell and R. K. Norris, *Rev. React. Org. React.*, 1, 65 (1972).

177. V. Massey, G. Palmer and D. Ballou, in *Oxidases and Related Systems*, T. E. King, H. S. Mason, and M. Morrison, Eds., University Park Press, Baltimore, 1973, p. 25.

178. S. Y. Vaish and G. Tollin, *Bioenergetics*, 2, 33 (1971).

179. H. Hölstron, *Acta Phys. Pol.*, 26, 419 (1964).

180. J. Rabani and S. O. Nielsen, *J. Phys. Chem.*, 73, 3737 (1969).

181. V. Massey, G. Palmer and D. Ballou, in *Flavins and Flavoproteins*, H. Kamin Ed., University Park Press, Baltimore, 1971, pp. 349-362.

182. Q. H. Gibson and J. W. Hastings, *Biochem. J.*, 83, 368 (1962).

183. R. F. Williams and T. C. Bruice, unpublished work.

184. D. Ballou, G. Palmer, and V. Massey, *Biochem. Biophys. Res. Commun.*, 36, 898 (1969).

185. F. Ziegeuner and G. Gubitz, *Monatsh. Chem.*, 101, 1547 (1970).

186. M. Yamasaki and T. Yamano, *Biochem. Biophys. Res. Commun.*, 51, 612 (1973).

187. N. Wiberg, *Angew. Chem. Int. Ed. Engl.*, 7, 766 (1968).

188. A. N. Fletcher and C. A. Heller, *J. Phys. Chem.*, 71, 1507 (1967).

188a. T. W. Chang, R. F. Williams, and T. C. Bruice, *J. Am. Chem. Soc.*, (1976).

189. T. Spector and V. Massey, *J. Biol. Chem.*, 247, 5632 (1972).

190. V. Massey and H. Neujor, unpublished results.

191. J. W. Hastings, C. Balny, C. Le Peuch, and P. Douznu, *Proc. Natl. Acad. Sci. U.S.*, 70, 3468 (1973).

192. C. Kemol and T. C. Bruice. *Proc. Natl. Acad. Sci. (U.S.).* To be submitted

193. G. Mitchel and J. W. Hastings, *J. Biol. Chem.*, 244, 2572 (1969).

# CHEMICAL PROPERTIES OF THE PHOTOTRAP IN BACTERIAL PHOTOSYNTHESIS

P.A. LOACH

*Department of Biochemistry and Molecular Biology*
*Northwestern University*
*Evanston, Illinois*

## 1  INTRODUCTION

For many years photosynthesis has been described by eq. 1. From

$$CO_2 \; + \; H_2O \; \xrightarrow{+h\nu} \; (CH_2O) \; + \; O_2 \tag{1}$$

the energetics of this reaction, it can be shown that a minimum of 3 quanta of light (of a wavelength = 680 nm) are required to drive the reaction to the right. Hence it has long been known that such a process must consist of several steps and cannot be a simple photochemical reaction. Today's scientist would most likely place the number of individual photoacts required to produce one molecule of oxygen at 8, and the number of subsequent dark reactions required to form the products indicated by the equation at between 30 and 50.

Photosynthesis is perhaps best defined as the sum of those processes by which light energy is utilized to drive the synthesis of biomolecules required for cell growth. Many organisms that depend on photosynthesis for survival do not evolve molecular oxygen. Van Niel [1-5] introduced a more general formalism of photosynthesis (eq. 2). In oxygen-evolving organisms, such as algae and plants, $H_2A$

$$CO_2 \; + \; 2H_2A \; \xrightarrow{+h\nu} \; (CH_2O) \; + \; H_2O \; + \; 2A \tag{2}$$

is water. But in organisms such as the photosynthetic bacteria, $H_2A$ can be various substrates ranging from hydrogen sulfide to a variety of organic molecules, such as succinate. A feature of photosynthetic bacteria attractive to the experimentalist is their ability to carry out a considerably simpler form of photosynthesis, perhaps indicative of a primative ancestor to the oxygen-evolving organisms. Indeed, part of the machinery used by oxygen-evolving organisms is highly similar to that utilized by the photosynthetic bacteria. It then seems logical to conclude that a fruitful approach to understanding the more complicated processes of oxygen-evolving organisms would be to first study and sort out how photosynthesis works in the simplest of organisms, such as is perhaps represented by a photosynthetic bacterium like *Rodospirillum rubrum.* This is the approach we adopted in our laboratory some 12 years ago and are still pursuing at present.

The disciplines applicable to the study of photosynthesis are indeed broad. The phenomena of light-energy absorption and transfer among molecules fall into the realm of physics; sorting out the molecular events that occur upon trapping the light energy is within the province of the photochemist; following the molecular gyrations of macromolecules that result from trapping the light energy is the discipline of biochemistry; and the integration of the whole process with the life of the cell is the area of the cell biologist. One might also include the energy engineer of the future who may greatly profit from an in depth understanding of the primary events of photosynthesis. It is small wonder that persons from many varied disciplines have worked on one part or another of the problem. It is also perhaps understandable why communication between these persons has been so poor. In this chapter I have attempted to bring several of these areas together, hopefully to enable better communication among researchers already working on some niche within the field, and also to introduce the subject to a newcomer or the interested passerby.

Work in our laboratory has been focused on the primary photochemical events of photosynthesis in general and bacterial photosynthesis in particular. Our strategy involves a three-level approach. One level deals with characterization of the *in vivo* (intact) system, a second is concerned with the isolation of an active phototrap complex, and the third is focused on constructing appropriate synthetic models. Work from our laboratory, and many others, in each of these three areas has progressed to the point that convergence toward the correct answer seems guaranteed. This is an area where there are far more *good* experiments to be done than there are people (or funds) to do them. In view of the intimate relationship of this process to energy conversion and food supply substantially increased Federal support would seem to be in order.

Over the past 10 years, a rebirth of activity in porphyrin chemistry and the first successful approaches to isolation of membrane functional proteins have provided important results for understanding the primary photochemistry of photosynthesis. To be sure, the factual information on hand is only a small fraction of that needed to completely understand this important process. Even so, a consideration of the relevant data currently available from these several fields of study does place constraints on the possible mechanisms of photosynthesis and may provide some insight for further experimentation. The approach used in this chapter is to first review the original experiments that seemed to have added something new to the chemical characterization of the phototrap and then to take a somewhat distant view to see where the observations are leading us. An effort is made to present a reasonable sample of the experimental data on the basis of which a characteristic property of the phototrap is postulated so that the reader may draw his own conclusions as to the validity of the assumption. After laying the groundwork for our basic knowledge of phototrap chemistry, current suggestions about the mechanism of the primary event are considered with a view toward future experiments that may provide a rigorous test of the hypotheses.

Each year an average of three comprehensive reviews are published in areas that deal in some measure with the primary events of photosynthesis. Although there are significant redundancies in these reviews, the latter reflect the high activity and relatively rapid progress in this field. Many new results need to be added each year. Several such reviews, as well as a few excellent books covering more general aspects of photosynthesis are cited in the reference list at the end of the chapter. Research over the next 5 to 10 years should result in a comprehensive understanding of the primary events in photosynthesis and of the structure and function relationships in the photosynthetic membrane, particulary that of photosynthetic bacteria.

A discussion similar in part to that found here in Sections 2 and 4 for the free radical nature of the primary electron donor and acceptor species has been previously presented (34a).

## 1.1 Chronological Events of Bacterial Photosynthesis

The events of photosynthesis may be conveniently separated on a temporal scale. It is expedient to use abbreviations for many of the important components of photosynthesis. These are defined when they first occur and also are collectively defined at the end of the chapter. A schematic representation of bacterial photosynthesis is presented in Figure 1. Unlike oxygen-evolving organisms, the photosynthetic bacteria appear to have only one photosystem (particularly the primitive types like *R. rubrum*). The products of this simplest form of photosynthesis beneficial to the cell are membrane-associated phenomena that drive transport processes as well as the formation of ATP. The photosynthetic bacteria grow in a medium

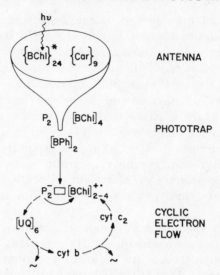

Figure 1. Schematic representation of bacterial photosynthesis. A photon of light ($h\nu$) is absorbed by the antenna pigments and rapidly transferred to the phototrap, resulting in the formation of an oxidized and reduced species. The components of the phototrap are indicated to be four bacteriochlorophyll (BChl), two bacteriopheophytin (BPh), and the primary electron acceptor $P_2$. Other abbriviations used are: Car, carotenoids; cyt $c_2$, cytochrome $c_2$; UQ, ubiquinone or coenzyme Q; cyt b, cytochrome b. When the trap is charged, the only species indicated are those where the oxidized entity and reduced entity are thought to reside. Arrows denote electron flow and the squiggle ($\sim$) which is produced from cyclic electron flow represents a high-energy intermediate or intermediate state, eventually giving rise to ATP.

containing a source of mild or strong reductant (e.g., succinate, $Na_2S$ thiosulfate, glucose, or glutamine). If they are growing in a medium with a reductant (e.g., $Na_2S$ and $CO_2$ as carbon source) insufficiently strong to drive the biosynthesis of cell components, they seem to utilize energy-linked reductions, such as reverse electron transport [6], to provide the needed reduced pyridine nucleotides. Thus the schematic representation of electron flow in photosynthetic bacteria is cyclic, that is, there is no direct accumulation of oxidant or reductant due to the photochemical event.

According to the chronological events of photosynthesis, certain parts of the process can be distinguished. These are given in Table 1 together with the time within which they are known to occur. After the absorption of a quantum of energy by bacteriochlorophyll or carotenoid, the energy migrates through the antenna pigments by some mechanism not yet clearly defined [9-14] and eventually results in excitation of the specific bacteriochlorophyll complex of

### TABLE 1  SEQUENTIAL EVENTS OF BACTERIAL PHOTOSYNTHESIS

| Event | Time (sec) | References |
|---|---|---|
| Light absorption by antenna pigments | $10^{-15}$ | 7, 8 |
| At least 100 transfers of excitation energy between antenna pigments | $10^{-12}$ | 9-14 |
| Conversion of singlet exited state to? | $7 \times 10^{-12}$ | 15,241 |
| Fluorescence observed from *in vivo* system | $5 \times 10^{-10}$ | 16-18 |
| Fluorescence of isolated Chl in organic solvent | $3 \times 10^{-8}$ | 19 |
| First photochemistry | $< 1 \times 10^{-6}$ | 20-22 |
| Most rapid secondary electron transport | $1 \times 10^{-6}$ | 22 |
| Time for complete cycle of electron transport | $10^{-1}$ | — |
| Half-time for direct return of electron from $P_2^-$ to $[BChl]_{2\text{-}4}^{+\bullet}$ (at room temperature) | $7 \times 10^{-2}$ | 23, 24 |

the phototrap $[BChl]_4$. Recent experiments [15, 241] provide evidence that within 7 psec the singlet excitation energy is converted either into another excited state (e.g., a triplet) or into the first photochemical products. Although the scheme of Figure 1 depicts only one photoreceptor complex (or photosynthetic unit), other such complexes are located nearby in the photosynthetic membrane so that excitation energy may be transfered to a significant extent between such complexes. In *R. rubrum,* under normal photosynthesis conditions only about 2% of absorbed light is lost by reemission as fluorescence. Some additional energy is lost by a delayed light emission process, but this is not well quantitated and depends greatly on the metabolic state of the organism. The first photochemically stable species are produced within 1 $\mu$sec, according to measurements [20-22] showing the characteristic absorbance changes for $[BChl]_{2\text{-}4}^{+\bullet}$ formation, which is described later. Subsequent electron transport then ensues, the first event of which may be very fast (e.g., 1 $\mu$sec). Good evidence exists (21, 22, 25) for assigning a cytochrome c type of molecule as the secondary electron donor species (eq. 3). It is important for this event to be significantly faster

$$[BChl]_{2\text{-}4}^{+\bullet} + Fe^{2+}cyt\ c_2 \rightleftharpoons [BChl]_4 + Fe^{3+}cyt\ c_2 \qquad (3)$$

than the recombination time (which is about 70 msec) of the oxidant and reductant produced in the primary photochemical event so that the "trapped" energy may not be wasted.

As indicated in Figure 1, the ratio of the number of antenna bacteriochlorophyll molecules to the number of phototraps was found to be about 24 in *Rhodospirillum rubrum,* one of the simplest known photosynthetic bacteria. The corresponding number for carotenoids is 9. The phototrap is considered to consist of the primary electron donor unit $[BChl]_4$, the primary electron acceptor species $P_2$, two bacteriopheophytin molecules $[BPh]_2$, whose role is not clearly understood, and an appropriate protein that specifically binds these components. The structure of bacteriochlorophyll is indicated in Figure 2. Bacteriopheophytin has the same structure as bacteriochlorophyll except the magnesium is replaced by two hydrogens, one on each of two pyrolle nitrogens. As is seen later there is good evidence that the phototrap and antenna complex exist together as

CHLOROPHYLL <u>a</u>

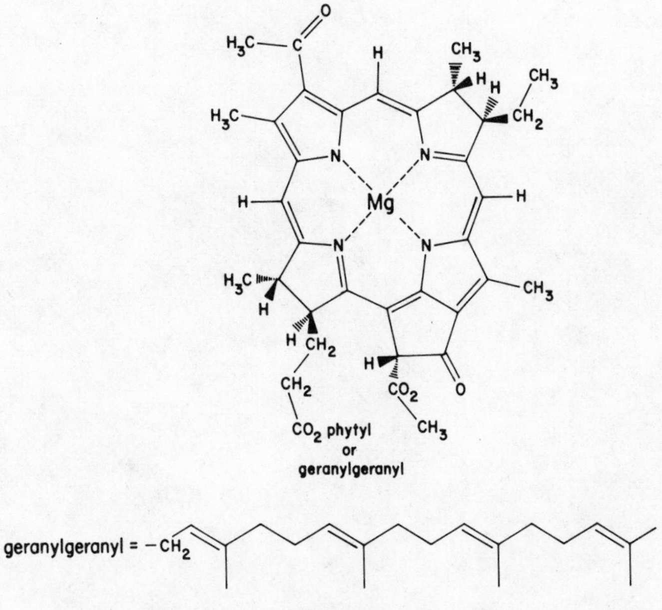

BACTERIOCHLOROPHYLL a

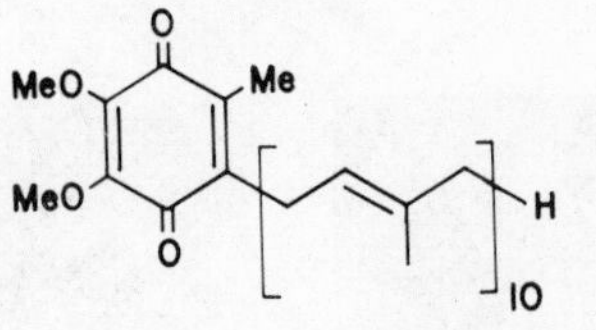

UBIQUINONE (CoQ₁₀)

Figure 2.   Structural formulas of phototrap components.

a morphological unit of molecular weight of about 125,000.

Secondary electron transport, and possibly the primary photo-chemical event itself, give rise to local and/or transmembrane ion and charge gradients [26-28]. These gradients are represented by the squiggle (~) in the scheme of Figure 1 and are viewed as coupling sites for the formation of ATP from ADP and $P_i$ catalyzed by a specific enzyme system [29]. Therefore energy released as a result of oxidation-reduction reactions is converted into anhydride bond formation in ATP, a process referred to as photophosphorylation.

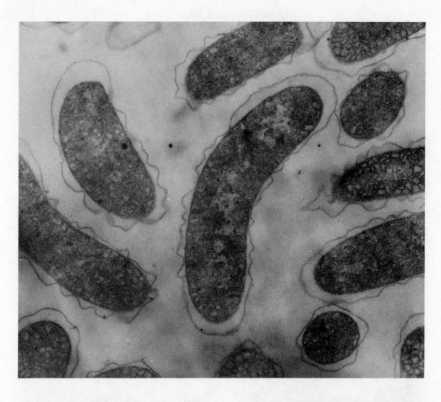

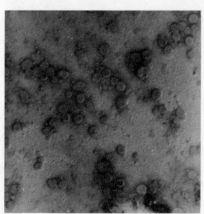 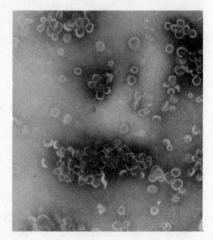

Figure 3. Electron micrograph of *R. rubrum* whole cells (top) and membrane vesicles (bottom). In all cases the magnification was 48,750 X. The membrane vesicles (chromatophores) in the lower left were shadowed with alumina and those shown in the lower right were negatively stained with phosphotungstate.

An intimate relationship exists between the electron transport components and the membrane, since an asymmetric electron and hydrogen ion flow is necessary to provide the driving force for phosphorylation and transport. It is not known whether more direct coupling of oxidation-reduction reactions to phosphorylation also occurs (in a manner similar to the process called substrate level phosphorylation (30)); the required intermediates for this process have not been demonstrated in spite of much effort expended in this direction. All those components that play a role in the events between light absorption and phosphorylation are either a part of the membrane structure or are bound to it. Figure 3 shows an electron micrograph of photosynthetic bacteria. Also shown in the figure are typical membrane particles obtained from whole cells by sonic oscillation, a process that fractures the membrane into smaller fragments called chromatophores. Complete photosynthetic activity, including photophosphorylation, has been demonstrated in chromatophores [31, 32]. In order to understand the chemistry possible in these systems, it must be kept in mind that the core of the membrane is a pseudo-fluid [33, 34], nonpolar, anhydrous phase, whereas the aqueous interfaces are highly polar. The possibility of nonpolar regions within the proteins, which may or may not be hydrated, and the possibility of isolated ion-pair regions within a hydrophobic matrix further complicate the oxidation-reduction reactions that may occur. In addition, transport of some molecules or groups between nonpolar and polar phases is probably an intimate part of these early events of photosynthesis.

## 2  THE PRIMARY ELECTRON-DONOR UNIT

### 2.1  Role of Bacteriochlorophyll

*2.1.1 In vivo* **Characterization.** Understanding the primary photochemical events of photosynthesis has depended greatly on two instrumental developments. One of these evolved from the thesis work of Duysens [35], in which he used difference spectroscopy to measure small light-induced absorbance changes in photosynthetic material. Subsequent application and extension of this technique by Duysens and others [36-40], together with the application [41-43] of flash photolysis [44-46] and pulsed laser methods [20, 21], led to the direct observation of the primary photochemical events. A recent measurement [47] of similar changes is shown in Figure 4. Based on a study of the effects of chemical oxidants on photo-

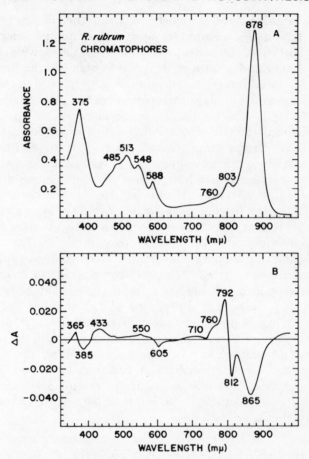

Figure 4. Absorbance (*A*) and light-induced absorbance change (*B*) of *R. rubrum* chromatophores (Taken from [47].

synthetic bacteria, Goedheer [48] suggested that the absorbance changes were due to photooxidation of a specific bacteriochlorophyll molecule. In the meantime, Kok, in a study of fragmented particles from chloroplasts of algae and spinach, was successful in partially "unmasking" the phototrap by acetone extraction of much of the antenna chlorophyll [49]. Working with these phototrap preparations enriched relative to antenna chlorophyll, Kok quantitated the light-induced absorbance changes and demonstrated that the same absorbance changes could be produced in the dark by adding an oxidant [e.g., $K_3Fe(CN)_6$] to the preparation. He also determined that the midpoint potential, $E_0'$, for the transition was 0.43 V. For this determination, the light-induced absorbance change at 700 nm was measured as the environmental potential was systematically varied. The logic utilized by Kok is shown by Scheme 1, in which

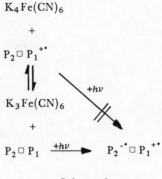

Scheme 1

$P_1$ and $P_2$ are defined as the primary electron donor and the primary electron-acceptor species, respectively. All other components of the phototrap and the light harvesting antenna are represented by the box. $K_3Fe(CN)_6$ and $K_4Fe(CN)_6$ are present in excess and effectively poise the redox potential of the system. Thus, as the environmental potential is raised sufficiently to cause oxidation of the primary electron-donor species, light-induced absorbance changes decrease because the donor has already lost an electron to the redox buffer. By systematically increasing the potential of the system, a midpoint potential of the $P_1^+/P_1$ equilibrium may be determined. Kok further demonstrated the reversibility of the transition by restoring the full light-induced absorbance change at low potential after the change had been quenched at high potential. These experiments were the forerunners of much subsequent work both in the optical "unmasking" of phototraps and also in using controlled potential techniques to measure *in vivo* oxidation-reduction reactions.

The second instrumental technique that has greatly increased our understanding of the primary photochemistry of photosynthesis is esr spectroscopy. This technique was first applied to the study of photosynthetic systems by Commoner et al. [50] and by Sogo et al. [51]. An extension of the esr type of measurement into a more recent technique, endor spectroscopy [52], has also been applied successfully [53, 54] to the study of the primary photochemistry of photosynthesis. The results suggest that both the absorbance and esr

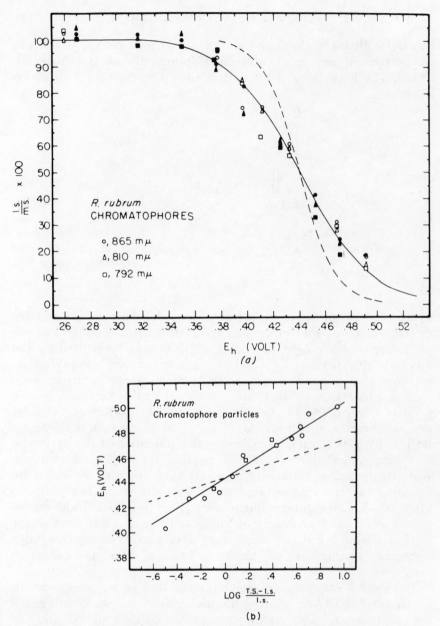

Figure 5. Potential dependency of the light-induced absorbance change (*a*) or esr signal (*b*) solid line and curve. The expected behavior for a one-electron reaction; dashed line and curve expected behavior for a two-electron reaction. A midpoint potential of +0.44 Volt was obtained by both procedure. In part *a* the ordinate is equivalent to the percent reduced, and in part *b* the abscissa is the same as log Ox/Red.

changes are directly related to the photooxidation of chlorophyll *a* in the phototrap.

A similar correspondence of midpoint potentials of light-induced absorbance change and light-induced esr measurements in photosynthetic bacteria has been demonstrated by Loach et al. [47, 55] and Kuntz et al. [38]. Typical data are shown in Figure 5.

*2.1.2* **Comparison with Model Compounds.** A definitive explanation of the light-induced absorbance changes and the light-induced esr signal could not be offered until the careful model studies on porphyrin oxidation carried out by Mauzerall and Fuhrop [56, 57] and by Felton et al. (58-60). Although earlier experiments dealt with the chemical oxidation of chlorophyll [61, 62] and bacteriochlorophyll [48], they were too qualitative to allow an identification of the higher oxidation state or, indeed, to determine whether or not these molecules were really reversibly oxidized. Fuhrop and

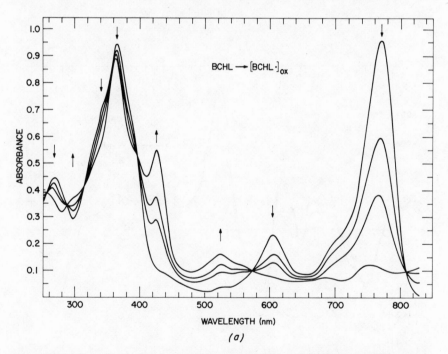

Figure 6. Changes in absorbance of bacteriochlorophyll in methanol caused by oxidation with iodine (*a*). An oxidized-reduced difference spectrum is plotted (*b*) using the data shown in *a*. For comparison, the light-induced absorbance changes observed in chromatophores of *R. rubrum* and *R. spheroides* are shown (*c*). (Taken from [38] and [55].)

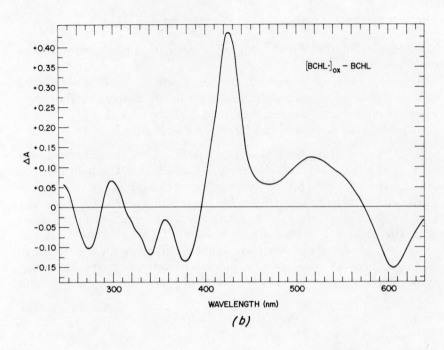

*(b)*

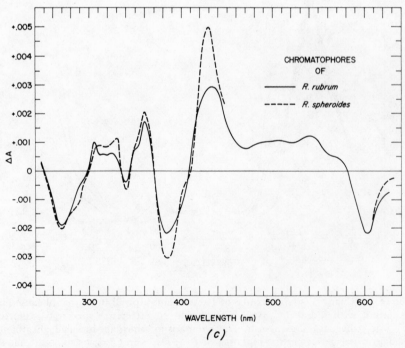

*(c)*

Figure 6. Continued.

Mauzerall showed [56, 57], however, that a comparison could be made between the light-induced absorbance changes observed in systems *in vivo* and those changes expected for reversible oxidation of bacteriochlorophyll. Such a comparison is shown in Figure 6. The overall similarity is striking and one cannot help but conclude that nearly all the light-induced absorbance changes observed in these bacterial systems are due to the photooxidation of the primary electron donor bacteriochlorophyll complex. It should be noted that the long wavelength band of bacteriochlorophyll is not shown in the comparison because it is so greatly red shifted in the *in vivo* system relative to bacteriochlorophyll dissolved in organic solvents.

This red shift of the bacteriochlorophyll band (monomeric BChl in methanol has $\lambda_{max}$ = 770 nm) both in the antenna ($\lambda_{max}$ = 880 nm in *R. rubrum*) and in the phototrap ($\lambda_{max}$ = 865 nm in *R. rubrum*) may probably be attributed to interaction between bacteriochlorophyll molecules and not to interactions between bacteriochlorophyll and protein or bacteriochlorophyll and phospholipid. The very elegant and careful work of J. J. Katz and collaborators [63-73] has shown that only hydrated (or in the presence of other bifunctional ligands that allow aggregation similar to that observed with water) oligomers of chlorophyll and bacterichlorophyll exhibit extensive red shifts of the long wavelength absorbance band. Water apparently serves as a bifunctional ligand, binding to two chlorophyll molecules and stabilizing an oligomeric species where significant $\pi$-overlap between the two molecules may occur (see Fig. 7C). Katz has further shown that in the absence of water, or any other bifunctional ligand, oligomer formation also occurs in solvents of poor nucleophilicity (e.g., aliphatic hydrocarbons) because the fifth coordination position of one chlorophyll ring can be occupied by a carbonyl group of the isocyclic ring of another chlorophyll molecule. However, this interaction can only occur readily if there is an appreciable angle between the chlorophyll rings (see Fig. 7B). Under these conditions $\pi$-overlap is poor and there is little shift in $\lambda_{max}$. It is of interest that bifunctional ligands other than water (e.g., amines, alcohols) may also allow significant $\pi$-overlap in dimeric species and result in some red shift of the $\lambda_{max}$ [72, 73, 152].

An alternate suggestion for the structure of the red-shifted oligomers (which has also been considered by Katz [110]) is presented here in Figure 8. Implicit in this model is the assumption that the $Mg^{2+}$ is pentacoordinated (and not hexacoordinated), because it is out of the center of the porphyrin ring by some 0.5 Å [232]. This

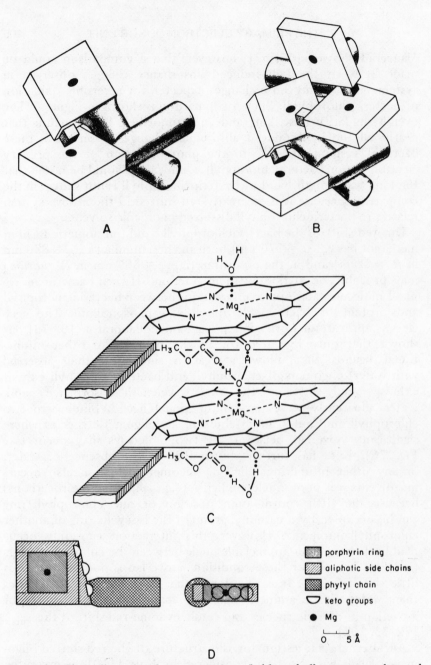

Figure 7. Three-dimensional representation of chlorophyll aggregates observed in organic solvents [72]: (A) dimer; (B) tetramer, or fragment of an oligomer; (C) the chlorophyll-water adduct, showing one unit of the (Chl · H₂O)$_n$ micelle; (D) dimensions of the chlorophyll molecule taken from a Courtauld space-filling model (Ballschmiter and Katz, 1969). (Taken from [73].)

106

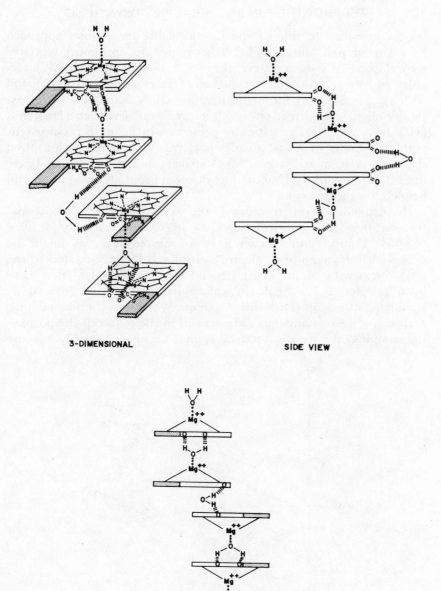

**3-DIMENSIONAL**

**SIDE VIEW**

**FRONT VIEW**

Figure 8. Alternative three-dimensional representation of chlorophyll aggregation in the presence of a bifunctional ligand.

107

leaves the opposite side of the ring available for a closer approach by a similar pair and would, I believe, give rise to a more effective $\pi$-overlap. As is seen later, the four bacteriochlorophyll molecules of the phototrap have sufficient interaction to give a band with $\lambda_{max}$ at 870 nm; one might therefore safely assume that more extensive oligomeric formation is not required to achieve such a large shift (100 nm) in $\lambda_{max}$. The structure presented in Figure 8 is consistent with the results of x-ray powder patterns for hydrated chlorophyll that always seem to have a 3.5 Å spacing in addition to the larger ones. Careful x-ray diffraction work on appropriate crystals should help clarify this question.

The dependency on structure of the oxidation-reduction potential for $\pi$-cation radical formation has been systematically studied [56-60]. Fuhrop and Mauzerall demonstrated that, as might be expected, this measure of electron affinity is directly related to the electronegativity of the central metal (see Fig. 9). $Mg^{2+}$ may have been selected for its role in chlorophyll in part because of its stabilizing effect on the $\pi$-cation radical. When transition metal ions, such as $Ni^{2+}$, $Co^{2+}$, and $Fe^{2+}$, are present in the center of the porphyrin complex, the $\pi$-cation radical cannot be formed readily because

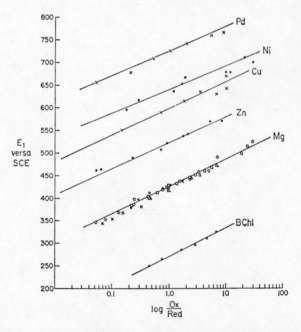

Figure 9.  Higher oxidation states of metalloporphyrins [56].

a higher oxidation state of the transition metal is more readily reached (see Fig. 10) [74]. Other properties of $Mg^{2+}$ that may have led to its selection for its role in chlorophyll are (a) $Mg^{2+}$ is diamagnetic and results in a highly fluorescent porphyrin complex—this is important to the role of chlorophyll as an antenna pigment and (b) it readily coordinates with oxygen- and nitrogen-containing ligands, which may be an important mode of binding both in antenna complexes and in the primary electron donor unit.

The questions might also be posed as to why a dihydroporphyrin (chlorophyll) or a tetrahydroporphyrin (bacteriochlorophyll) was selected as the oxidation level of the porphyrin and why a fifth ring, the so-called isocyclic ring, was selected [72, 73, 75]. The first

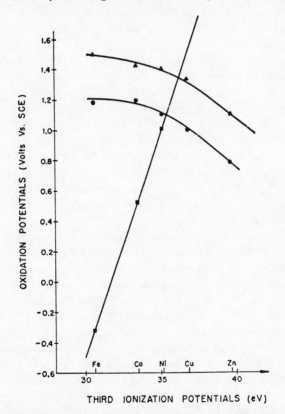

Figure 10. Metalloporphyrin redox properties compared with the ionization potentials of the metal [74a]. Oxidation potentials of the [M(II)TPP] complexes plotted against third ionization potentials of the metal atoms (■) metal valency change; (●) first ligand oxidation; and (▲) second ligand oxidation. (Taken from [74].)

question is readily answered: dihydroporphyrins and tetrahydropor-
phyrins have an intense red or far red absorbance band that allows
them to absorb light very effectively at wavelengths near the maximum of the sun's radiant flux reaching the surface of the earth. The
role of the fifth ring, on the other hand, could be attributed to its
usefulness in specific aggregation, which has assured high efficiency
for energy transfer in antenna chlorophyll and bacteriochlorophyll.
As discussed later, its role in special pair formation may also be
essential for the functioning of the primary electron donor unit.
Mauzerall has made the novel suggestion that the fifth ring is directly
involved in oxygen evolution [76]. However, experiments with *in
vivo* systems so far do not support this proposal. The phytol (or
geranyl-geranyl [77]) tail seems to provide a hydrophobic anchor,
which is probably important for binding to protein and stabilizing
the complex in the membrane.

Model studies on porphyrin derivatives have also contributed
greatly to understanding the esr signal observed in photosynthetic
systems. As mentioned above, the esr signal was early assigned to the
oxidized primary electron donor species because its redox properties
matched those observed in the quenching of the light-induced
absorbance changes in the bacteriochlorophyll region [55]. The
relationship between the light-induced absorbance and esr changes in
bacterial systems has been further documented by comparison of the
decay kinetics [78] and the rise and decay kinetics [79, 80] of the
two changes under a variety of conditions. An exact kinetic match
was observed in all these studies.

Of those methods applied to the study of photosynthetic systems,
isotope substitution proved to be one of the most useful. Kohl et al.
[81] grew algae and photosynthetic bacteria on deuterium-containing culture media and examined the photoinduced esr signals in these
species. A two-to three fold narrowing of these signals was observed.
McElroy et al. [79] conducted similar studies in which constant
proportionality was shown to exist between the amount of isotopic
narrowing of the photoproduced signal in chromatophores and the
*in vitro* bacteriochlorophyll cation radical. Recent extension of these
techniques by Katz and co-workers [72, 73, 82, 83] underscores the
great potential of the method. By a systematic use of carbon,
nitrogen, and hydrogen isotopes, much can be learned about the
spin distribution in the cation radical and about chlorophyll-chloro-
phyll (or BChl-BChl) and chlorophyll-protein (or BChl-protein)
interactions in the phototrap.

*2.1.3* **Other Properties of the Phototrap.** One of the early criteria used in deciding whether the light-induced absorbance changes and esr signal were part of the primary photochemistry was to see if they would occur at very low temperatures. Since the primary event takes place in a pseudosolid environment it might still occur at low temperature, whereas any secondary electron transport that requires the movement of molecules or parts of molecules could not occur. Arnold and Clayton [37] demonstrated that the light-induced absorbance changes in chromatophores from *R. spheroides* could be produced at liquid helium temperatures and that they rapidly decayed in the dark. Andres et al. [84], and later McElroy et al. [79,80], demonstrated that the light-induced esr signal was also reversibly produced at low temperatures in bacterial chromatophores.

Quantum-yield measurements further substantiated that the light-induced esr and absorbance changes were reflecting a change in oxidation state of the same molecular species. Early measurements [85, 86] of quantum yields indicated that light trapping might occur with a near ideal efficiency. More precise measurements later confirmed that the primary electron donor species in photosynthetic bacteria became oxidized with a quantum yield near unity (e.g., 0.95 + 0.05 [25, 87-89]. These studies also demonstrated that when the concentrations of the species being oxidized were normalized to the total bacteriochlorophyll concentration, they were the same to within 5%, whether determined by esr or by absorbance-change measurement [87].

## 2.2 Cooperativity Between Bacteriochlorophyll Molecules in the Donor Unit

*2.2.1* **Optically Unmasking the Phototrap.** Evidence that several bacteriochlorophyll molecules act cooperatively in the primary electron donor unit was first acquired from the observation that these several molecules were necessary for phototrap activity in "unmasked" phototraps [55, 90]. The first experiments of this type successfully unmasked the phototrap by selective degradation of antenna pigments [38, 55, 90]. Figure 11 shows the absorbance spectra of such unmasked phototraps. The absorbance bands observed in the near infrared are now refered to as a typical reaction center spectrum. Using solvent extraction on these unmasked phototraps, Clayton demonstrated that the phototrap pigment was bacteriochlorophyll [91, 92].

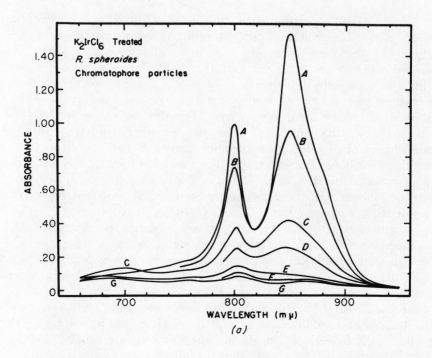

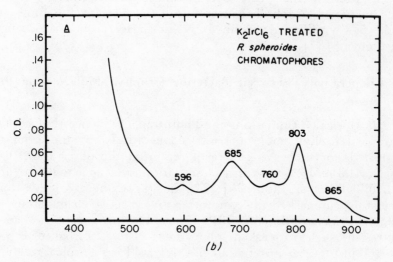

Figure 11. Absorbance spectrum of unmasked phototraps: (*a*) Degradation of antenna bacteriochlorophyll by successive $K_2IrCl_6$ treatments of chromatophores; (*b*) resultant spectrum of unmasked reaction center after $K_2IrCl_6$ treatment. (Taken from [38, 55].)

112

The application of detergents to selected membrane systems of photosynthetic bacteria has allowed further progress in isolation of small protein-pigment complexes that are still photochemically active. These reaction centers were first prepared by Reed and Clayton [93]. The best characterized preparations are from the R-26 carotenoidless mutant of *R. spheroides* [94-98]. The protein components of these reaction centers are discussed in detail later. According to spectral [99] and chromatographic properties [100, 101] the bacteriochlorophyll and bacteriopheophytin extracted from reaction center preparations appears to have a structure identical with bacteriochlorophyll *a* and bacteriopheophytin *a*.

**2.2.2 Circular Dichroism Measurements.** Interaction between the bacteriochlorophyll molecules in the donor unit was first demonstrated by circular dichroism measurements on reaction center preparations [102-104]. Dratz and co-workers [102] showed that at least three bacteriochlorophyll molecules interacted strongly with each other during light absorption. Clayton originally suggested [91] the existence of three interacting bacteriochlorophyll molecules in the phototrap. This suggestion was based on the analysis of the components present in unmasked phototraps and on the observation that part of the light-induced absorbance changes (those between 785 and 830 nm in Fig. 4) could be interpreted as a hypsochromic shift of the bacteriochlorophyll absorption. It was proposed that this shift arose from the close proximity to BChl molecules of a positive charge on the photooxidized primary electron-donor molecule during the primary photochemical event.

**2.2.3 Esr and Endor.** After it was shown that higher oxidation states of metalloporphyrins could be generated readily in organic solvents and after their spectral and redox properties were ascertained [56, 57], the esr properties of these radicals were carefully examined [58-60, 105]. Good agreement was found between the *g*-value of the cation radical of chlorophyll or bacteriochlorophyll in methanol and that of the primary electron-donor species *in vivo*. On the other hand, a difference was found consistently between the line widths of these two signals. In all cases the cation radical generated in solution had a wider signal than the respective *in vivo* species. The occurrence of these systematic differences in line width in all photosynthetic organisms has been documented by Norris et al. [106]. Hanna et al. [107] and Vincow and Johnson [108] showed that the second moment of an esr signal is related to the sum of the second moments of each hyperfine interaction. Using this concept, Norris et al. [106] derived an expression relating the line width of a one-electron oxidized aggregate of chlorophyll molecules to the number of molecules in the aggregate. These calculations were based on the assumptions that all the spectra used had symmetrical Gaussian line-shapes and that the unpaired $\pi$-electron was distributed uniformly over all the molecules in the aggregate. The results of these calculations clearly imply that the physical state of the primary electron donor may be a hydrated special pair ($BChl \cdot H_2O \cdot BChl$) or an equivalent dimer with some other bifunctional ligand. For an example of the way in which Katz has envisioned this structure, see Figure 7C. Fong has proposed a somewhat modified version of the special pair involving two $H_2O$ molecules [227].

An alternative to Katz's special pair model may be considered and has been previously alluded to (Fig. 8) in explaining the almost 100 nm red shift of the long wavelength band of the bacteriochlorophyll donor as compared with that of the monomeric bacteriochlorophyll in organic solvents. Indeed, from a consideration of the model studies of Katz and co-workers [72, 73], the absorbance spectra of unmasked phototraps and reaction center preparations (Fig. 11) require that the four bacteriochlorophyll molecules present interact significantly. For the present purposes, the model of Figure 8 is unique in that the two strongly interacting bacteriochlorophyll molecules do not have a water molecule between them and can therefore approach each other more closely, allowing for extensive charge transfer interaction. Appropriate groups in the protein and/or the two bacteriopheophytin molecules may play a role in hydrogen bonding in place of water. With respect to the possible role of bacteriopheophytin, it is interesting to note that Norris et al. [72] showed a specific molecular complex of two molecules of chlorophyll $a$ and one of pheophytin $a$ to be very stable. However, a similar stable species has not yet been demonstrated for bacteriochlorophyll and bacteriopheophytin.

Recently the technique of endor spectroscopy has been used independently by Norris et al. [53, 231] and Feher et al. [54] to probe the structure of bacteriochlorophyll in the donor unit. Endor spectroscopy is extremely useful for investigating the hyperfine structure of radical species. It is, however, limited by the fact that the sensitivity of endor spectrometers is several orders of magnitude less than that of esr spectrometers, and frequently one must employ a very low temperature to achieve acceptable sensitivity. The *in vivo* and *in vitro* endor spectra of bacteriochlorophyll are shown in Figure 12. Two distinct splittings are observed in both *in vivo* and *in vitro* samples; these peaks have been assigned to the coupling of methyl groups on the porphyrin ring of the bacteriochlorophyll. There is a

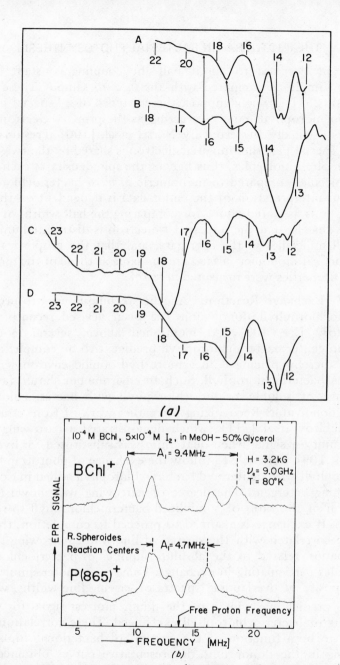

Figure 12a. ENDOR SPECTRA. a.) (A) $^1$H Chl $a$ in $C^2H_2Cl_2-C^2H_3O^2H$ (4:1) oxidized by $I_2$. Frequency scale (MHz) compressed by a factor of 2 ($T = 108°C$); (B) S. lividus oxidized by $K_3Fe(CN)_6$. Coupling constants are compared at points connected by lines ($T$) = 108°C); (C) BChl in $C^2H_2Cl_2-C^2H_3O^2H$ (4:1) ($T = 15°K$); (D) R. rubrum oxidized by $K_3Fe-(CN)_6$ ($T = 15°K$). (Taken from [231] and [54].)

115

significant decrease to about half the coupling constant for the *in vivo* sample as compared with the *in vitro* sample. These results [53, 54, 231] strongly support the existence of a "special pair" of bacteriochlorophyll molecules acting as the primary electron donor unit, as originally suggested by Norris et al. [106]. Presumably, in such a special pair the unpaired electron is shared by the two bacteriochlorophyll molecules, thus halving the spin density at each site on the molecules compared to monomeric *in vitro* bacteriochlorophyll. Such an interpretation of the endor data is dependent on the same assumptions as stated above for comparing the half widths of the esr signals, that is, that the unpaired $\pi$-electron is distributed uniformly over all molecules in the aggregate, and that the properties of the hypothetical monomer *in vivo* are the same as those of the monomer whose properties were measured *in vitro*.

2.2.4 **Exchange Reaction.** Another technique used to probe the bacteriochlorophyll donor unit has been devised recently in my laboratory [24, 109]. As mentioned above, several models of reaction center bacteriochlorophyll predict it to be complexed with water, thereby placing it in a more hydrophilic environment than antenna bacteriochlorophyll. Such an environment should favor the exchange of soluble bacteriochlorophyll with bacteriochlorophyll in the donor unit. Recognizing the earlier work of Kohl et al. [81] and McElroy et al [111], which demonstrated a narrowing of the donor unit esr signal when deuterium was substituted for hydrogen, we [24, 109] were able to follow the exchange of phototrap bacteriochlorophyll with deuterated bacteriochlorophyll added in solutions containing detergent. The degree of narrowing was shown to be a function of the ratio of solubilized bacteriochlorophyll to reaction centers. If exchange is assumed to proceed to completion, this ratio can be correlated with the degree of line width narrowing to yield information related to the absolute number of bacteriochlorophyll molecules participating in exchange. Using a computer simulation of a composite of overlapping spectra of varying line widths, we were able to predict line widths of the donor unit esr signal for various amounts of bacteriochlorophyll exchanged. These correlations were shown to be a function of the amount of bacteriochlorophyll exchanging in the donor unit. Representative curves obtained form these simulations are shown in Figure 13. Experimental data are also shown in this figure; these data best fit a model in which four phototrap bacteriochlorophyll molecules are exchanged and all four of these share the unpaired electron when the donor unit is oxidized. We also have shown that this exchange is reversible and

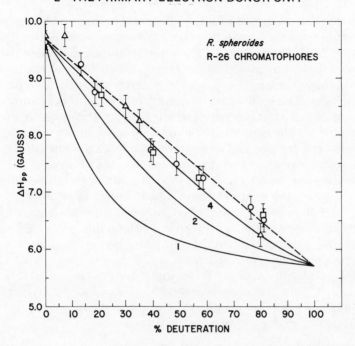

Figure 13. Variation of the peak-to-peak esr line width ($\Delta H$pp) of the *in vivo* bacteriochlorophyll radical as a function of the total percent of exchangeable perdeuterobacteriochlorophyll in solution. Solid lines represent theoretical variation for 1, 2, 4, and an infinite number of exchangeable bacteriochlorophyll molecules in the bacterium's reaction center. Empirical data is given for the hydrogen-containing *in vivo* system (○) and for the perdeuterated *in vivo* system (Δ).

have obtained verifying quantitative data by isolating [14]C-bacterio-chlorophyll displaced from the phototrap.

There is some uncertainty associated with comparing the experimental points with the theoretical curves of Figure 13, since the latter are based on the assumption that all trap sites react with the added bacteriochlorophyll. If for example 25% of the traps were simply not reactive, the appropriately corrected theoretical curve for two molecules sharing the unpaired electron under this condition would significantly overlap the theoretical case for all four molecules sharing the unpaired electron when all sites react. According to exchange experiments conducted with deuterated *R. spheroides* R-26 chromatophores, less than 5% of the phototraps could have been unreactive for those experiments. Thus the criterion of full reactivity of all phototraps is met if it can be assumed that chromatophores prepared form cells grown on the normal hydrogen-containing medium behave like those grown on deuterium-containing medium. Even so, because of these uncertainties the endor data are generally viewed as providing the stronger argument for the unpaired electron being shared by a special pair rather than four bacteriochlorophylls. The endor data, however, suffers from the shortcoming that it is obtained at low temperature ($16°$ K); the fact that the dark decay time for the light excited phototrap is five times faster at these low temperatures (see Figure 22) than at $268°$ K suggests that a significant rearrangement in phototrap structure has occured at the lower temperature. However I have left this question open in the scheme of Figure 1 where the oxidized donor unit is shown as $[BChl]_{2-4}^{+\cdot}$.

The exchange reaction described above has the potential of being an extremely useful tool in research on the binding site and properties of the primary electron donor unit. We have initiated such a study by observing the ability of BChl to exchange *in vivo* with various porphyrin analogs. Preliminary results show that bacteriochlorophyll and alkaline-hydrolyzed bacteriochlorophyll readily exchange with the donor unit while bacteriopheophytin and chlorophyll *a* do not exchange with bacterial systems.

**2.2.5  A Triplet State of BChl In Vivo.**  Since it is the photoexcited bacteriochlorophyll donor unit that initiates electron transport, the identity of the excited state that precedes the electron transport has received much attention recently. Using *Chromatium D* chromatophores, Leigh and co-workers [112, 113] were able to demonstrate the existence of a triplet state at liquid helium temperatures. The triplet signal could only be observed when the primary electron-acceptor species was reduced, either by light or chemically. The spectrum assigned to the primary electron donor is shown in Figure 14. Comparison of the properties determined by esr of this *in vivo* triplet state to those observed *in vitro* should provide further information about the organization and interaction of bacteriochlorophyll molecules in the donor unit, even if the triplet state is not an important intermediate in photosynthesis. In this regard its relevance may be limited by the fact that it can be observed only at very low temperature and in a reduced state of the phototrap. A recent comparison of *in vivo* and *in vitro* triplet states as measured by esr provides evidence for the expected involvement of more than one bacteriochlorophyll molecule in the *in vivo* triplet state [233].

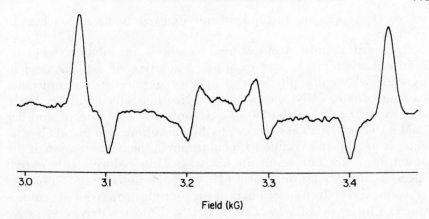

Figure 14. Light-induced esr spectrum of the triplet state of reaction center bacteriochlorophyl in *Chromatium* D chromatophores as observed by Leigh and Dutton [113].

## 2.3  Photochemical Model Studies Utilizing Porphyrins

Chlorophyll and porphyrins in general have long been known to photosensitize oxidation-reduction reactions in solution [19, 114, 115]. According to the published literature, the term photosensitization does not imply a mechanism, but rather conveys the fact that oxidized and/or reduced compounds may be demonstrated in a particular solution after light absorption by a porphyrin analog. Often there is no apparent change in the oxidation state of the photosensitizer. Some of the earliest demonstrated reactions of this type were described by Krasnovskii [114].

As reviewed by Seely [19], chlorophyll photochemistry in solution can be described as following one of the two paths shown by eq. 4.

$$Chl + h\nu \rightarrow Chl' \text{ (singlet excited state)} \qquad (4)$$

$$Chl' \rightarrow Chl* \text{ (triplet excited state)}$$

$$(A) \quad Chl' \text{ or } Chl* + O_x \rightarrow Chl^{+\bullet} + O_x^{-\bullet}$$

$$Chl^{+\bullet} + Red \rightarrow Chl + Red^{+\bullet}$$

$$(B) \quad Chl' \text{ or } Chl* + Red + H^+ \rightarrow ChlH^\bullet + Red^{+\bullet}$$

$$ChlH^\bullet + O_x \rightarrow Chl + O_x^{-\bullet} + H^+$$

where $O_x$ represents an oxidant and Red represents a reductant in solution.

The most readily demonstrated reactions in solution seem to follow mechanism B. For example, a solution of ascorbic acid in pyridine with chlorophyll *a* can be used to photoreduce molecules like ferredoxin, FMN, and pyridine nucleotides, albeit with relatively low quantum yields (0.02-0.07). If only ascorbic acid or another mild reductant, such as cysteine, hydrogen sulfide, and phenyl hydrazine, is present in pyridine with chlorophyll, then illumination in the absence of air can result in a two-electron-reduced chlorophyll species, probably an isophlorin [19, 116]. Presumably such a species is formed by disproportionation of one-electron-reduced chlorophyll: $ChlH\cdot + ChlH\cdot \rightleftharpoons Chl + ChlH_2$. Unfortunately, conclusive evidence for a role of one-electron-reduced chlorophyll has not been experimentally obtained.

More relevant to the apparent role of chlorophyll and bacteriochlorophyll in photosynthesis is the general reaction mechanism A in eq. 4. Unfortunately again, such reactions are even less well characterized mechanistically than those that seem to proceed by mechanism B. Of particular interest to photosynthesis, quinones can be efficiently photoreduced in alcoholic solvents at room temperature in the presence of a porphyrin analog in the absence of air [117, 118]. This is an interesting reaction but not especially impressive, since alcohol reduction of quinones (to the hydroquinone) is a thermodynamically spontaneous reaction and readily occurs if the excitation energy is provided by direct absorbance of near uv light by the quinone species. Harbour and Tollin [119] have recently characterized a model system in which bacteriochlorophyll donates an electron to quinone in acetone at $-100°C$. Under these conditions they were able to demonstrate by esr measurement the $\pi$-cation radical formation of bacteriochlorophyll. It is obvious that a great deal of research needs to be conducted with photochemical model systems that have more relevance to photosynthesis. In a sense model systems may prove somewhat more difficult to sort out than the *in vivo* systems. In spite of their small size, model photochemical systems usually are able to undergo many more possible reactions than a phototrap, in which the possibilities have been restricted by selection during about a billion years of trial and error.

Because it is probably not possible to prepare unique hydrated dimer, trimer, or tetramer aggregates of chlorophyll or bacteriochlorophyll in any solvent system, and since these would clearly be the most relevant models for understanding the chemistry and photochemistry of the primary electron donor unit, we have embarked on the synthesis of covalently linked porphyrins of the type shown by

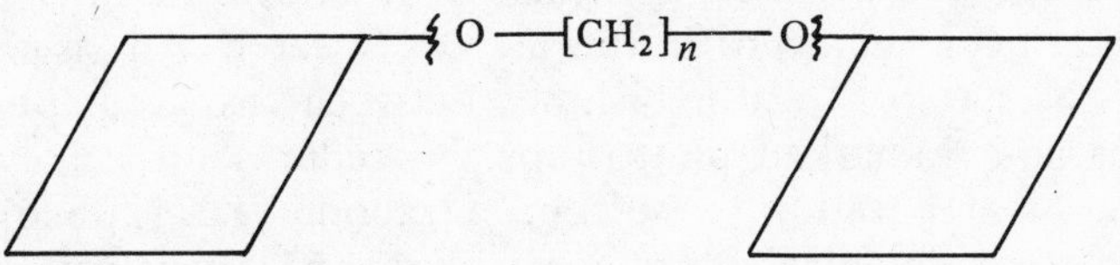

**I**

structure **I**, where the rhombus represents a molecule like that shown in structure **II**. Dolphin and  co-workers [120] have previously pre-

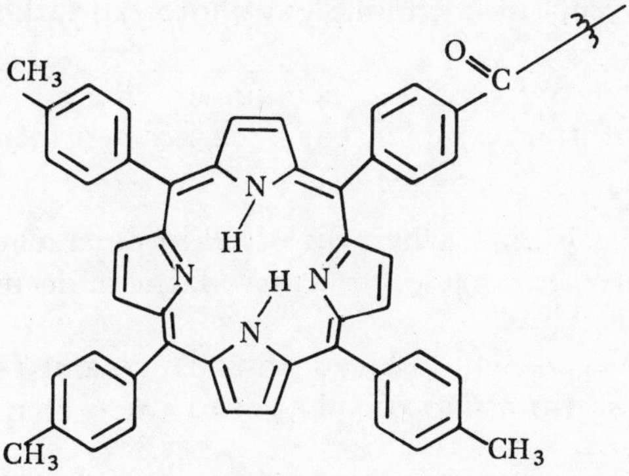

**II**

pared a porphyrin dimer using amide linkages to the bridging group and have studied energy transfer between the porphyrin molecules. We have recently [121-123] been able to prepare dimer, trimer, and tetramer species and are now ready to systematically examine the chemical and photochemical properties of each of these analogs as the value in $n$ varies from 2 to 12 and when the individual porphyrins contain a metal (e.g., $Mg^{2+}$) or are metal free. Clearly this type of model approach should be helpful in eventually understanding both energy transfer and the primary photochemical events.

## 3  THE PROTEIN COMPONENT OF PHOTOTRAP COMPLEXES

The goal of isolating from photosynthetic material a small (perhaps of mol wt = 50,000-150,000) protein-chlorophyll or protein-bacter-iochlorophyll complex that is still able to carry out the primary photochemical event has long been envisioned [90, 91, 124-131]. Early attempts to isolate such complexes yielded rather large (mol wt $\cong 10^6$) fractions, which still had attached to them many redox functional groups whose role was known to be associated with secondary electron transport. In fact, some preparations still seemed to be membrane vesicles from which an unknown number of components had been removed. In this area of research there is a need to set up criteria for distinguishing between isolated phototrap complexes and unmasked phototraps. Procedures for achieving the latter type of preparation by solvent extraction [127], pheophytinization [90], or chemical bleaching [38, 47, 55] have already been cited in this chapter. Unfortunately, many detergent treatments (e.g., sodium dodecyl sulfate), often utilized in the preparation of phototraps and reaction centers, also can cause selective solubilization which may result in an unmasked phototrap rather than a purified one.

The following are suggested as minimal criteria for establishing that an active phototrap complex has been isolated from photosynthetic material:

a.   Smallness of size when dissolved in detergent-containing solution. From preparations so far achieved, the molecular weight should be $\leq 125,000$.

b.   *In vivo* type light-induced absorbance and esr changes as expected to arise from photooxidation of the primary electron-donor unit.

c.   Quantum yield of $0.95 \pm 0.05$ for the primary photochemistry.

d.   Low-temperature formation and decay kinetics that agree with *in vivo* systems.

e.   Absence of phospholipid, although detergent molecules may be tightly bound in place of phospholipid.

f.   No appreciable transition metal content. This point will still be debated by some because of arguments regarding iron as a candidate for the primary electron-acceptor species, as is discussed in a later section. A requirement for iron could, in fact, vary from organism to organism.

g.   Minimal number of polypeptides. It is very difficult at this juncture to know how uniform this particular criterion will be. The minimal number found so far is 2 for a reaction center preparation from a carotenoidless mutant of *R. spheroides* [132] and perhaps 2 for a photoreceptor complex preparation from wild-type *R. rubrum* [133-135].

h.   High specificity and quantum yield for secondary electron transport.

i.   Unchanged absorbance and circular dichroism spectra of the reaction center.

j.  For those preparations that retain antenna pigments, mainten-ance of the *in vivo* $\lambda_{max}$, energy transfer, and fluorescence proper-ties.

k.  Unchanged redox properties for the primary electron-donor unit.

In practice all these criteria are seldom applied to any single pre-paration. After a few more years research it should be possible to add other more restrictive requirements with regard to the protein and the primary electron-acceptor species.

## 3.1  Isolation of Active Protein-Bacteriochlorophyll Complexes

Two distinct kinds of preparations have developed in parallel. The photoreceptor complex is an antenna-phototrap-protein complex, whereas the reaction center preparation is an antenna-free phototrap-protein complex.

*3.1.1* **Photoreceptor    Complexes.**    *3.1.1.1 Preparation    and Analyses.* This procedure [136, 137] combines three conditions commonly used for dissociating proteins, namely, alkaline solution (pH 11-12), high urea concentration (6 *M*), and 0.5 to 3.0% of a nonionic detergent (e.g., Triton X-100). A typical preparation uti-lizing these AUT (alkaline-urea-triton) conditions is outlined in Figure 15. The phototrap was shown to withstand these conditions with little apparent loss in activity or integrity. Utilization of this pro-cedure in our laboratory was shown to convert quantitatively all membranes studied (*R. rubrum, R. spheroides,* R-26 mutant of *R. spheroides, R. capsulata, Chromatium,* chloroplasts, and mitochon-dria) into components that have a particle size smaller than 150,000. For each membrane the pH and the amount of detergent required to achieve complete conversion varies; for example, *R. rubrum* chrom-atophores require pH = 12.0 and 3% Triton X-100, while for chloro-plasts, pH = 10.8 and 0.5% Triton X-100 gave satisfactory dissolu-tion.

After converting the membrane to solubilized detergent-lipid, protein-lipid, protein-detergent, and protein-pigment complexes, column electrophoresis was found to be a mild procedure that caused separation of many of the solubilized membrane components. The particular electrophoresis employed made use of a high concentra-tion of metal chelators in the form of the ampholytes used in electro-focusing procedures. The presence of such chelators was probably important for iron removal, which this step appears to accomplish, at

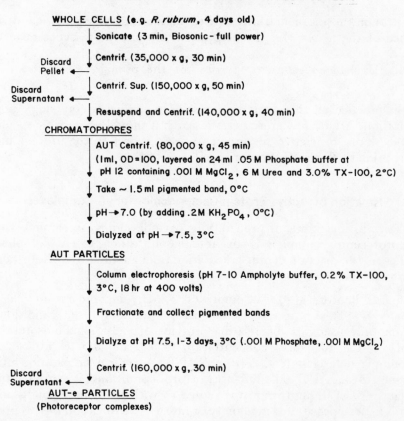

WHOLE CELLS (e.g. *R. rubrum*, 4 days old)

    Sonicate (3 min, Biosonic - full power)

Discard    Centrif. (35,000 x g, 30 min)
Pellet ◄—

Discard    Centrif. Sup. (150,000 x g, 50 min)
Supernatant ◄—

    Resuspend and Centrif. (140,000 x g, 40 min)

CHROMATOPHORES

    AUT Centrif. (80,000 x g, 45 min)
    (1ml, OD = 100, layered on 24 ml .05 M Phosphate buffer at
    pH 12 containing .001 M MgCl₂ , 6 M Urea and 3.0% TX-100, 2°C)

    Take ∼ 1.5 ml pigmented band, 0°C

    pH→7.0 (by adding .2M KH₂PO₄ , 0°C)

    Dialyzed at pH →7.5, 3°C

AUT PARTICLES

    Column electrophoresis (pH 7-10 Ampholyte buffer, 0.2% TX-100,
    3°C, 18 hr at 400 volts)

    Fractionate and collect pigmented bands

    Dialyze at pH 7.5, 1-3 days, 3°C (.001 M Phosphate, .001 M MgCl₂)

Discard    Centrif. (160,000 x g, 30 min)
Supernatant ◄—

AUT-e PARTICLES
(Photoreceptor complexes)

Figure 15. Flow diagram for preparation of photoreceptoreceptor complexes by the AUT-e method.

least with *R. rubrum* material [138]. Also present during the electrophoresis is 0.2% Triton X-100 which is used to retard reaggregation of the membrane components being separated. Analytical data for such a preparation from *R. rubrum* are summarized in Table 2. For such preparations, excellent phototrap activity has been demonstrated both at room and low temperature with a quantum yield of 0.95 [136]. Even though these preparations yield small particles, the latter still retain antenna bacteriochlorophyll and carotenoid pigments. The absorbance spectra of the purified photoreceptor complex showed that the environment of the antenna pigments is little changed from the *in vivo* state. Redox properties of the primary electron-donor unit are also unchanged relative to the *in vivo* state. Apparent redox equilibria involving the primary electron-acceptor

## TABLE 2 ANALYSES OF PHOTORECEPTOR COMPLEXES [133, 139]

| Component | R. Rubrum Mol Wt = 100,000 ± 25,000 [136] | R. Spheroides (wild type) | |
|---|---|---|---|
| | | Active Mol Wt = 100,000 ± 25,000 | Inactive Mol Wt = 100,000 ± 25,000 |
| Bacteriochlorophyll $a$[a] | 28 | 30 | 25 |
| Bacteriopheophytin $a$[a] | 2 | $\geqslant 2$ | — |
| Carotenoid[a] | 9 | 9 | 9 |
| P-lipid[a] | $\leqslant 1$ | $\leqslant 1$ | $\leqslant 1$ |
| Ubiquinone[a] | 3.5 [139] | 5 [139] | <0.6 [139] |
| Iron[a] | <0.3 | <1 | — |
| Copper[a] | <0.2 | — | — |
| Manganese[a] | <0.2 | — | — |
| $\Phi$ for P865 oxidation | 0.9 | — | — |
| $E'_0$ of P865 | 0.44 V | — | — |
| Polypeptides | 32,000[b] 16,000[b] 30,000[c] (24,000)[c] (21,000)[c] 9000-12,000[c] | 24,000[c] 21,000[c] 12,000[c] 9000[c] | (65,000)[c] (50,000)[c] (45,000)[c] (35,000)[c] (14,000)[c] 11,000[c] |

125

## TABLE 2 ANALYSES OF PHOTORECEPTOR COMPLEXES (Continued)

| Component | R. capsulata (140) Mol Wt = 100,000 ± 25,000 | R. Spheroides R-26 Active Mol Wt = 100,000 ± 25,000 | R. Spheroides R-26 Inactive |
|---|---|---|---|
| Bacteriochlorophyll $a^a$ | 75 | 4 | — |
| Bacteriopheophytin $a^a$ | — | 2 | — |
| Carotenoid$^a$ | 30 | 0 | 0 |
| P-lipid$^a$ | — | ≤1 | — |
| Ubiquinone$^a$ | — | — | — |
| Iron$^a$ | — | — | — |
| Copper$^a$ | — | — | — |
| Manganese$^a$ | — | — | — |
| Φ for P865 oxidation | — | — | — |
| $E'_0$ of P865 | — | — | — |
| Polypeptides | $(30,000)^c$ $((24,000))^c$ $((21,000))^c$ $12,000^c$ $9000^c$ | $24,000^c$ $21,000^c$ $(10,000))^c$ | $10,000^c$ |

$^a$ Estimated number per phototrap.
$^b$ Weber and Osborn method of SDS PAGE [141].
$^c$ Tris method of SDS PAGE [95, 96, 132]. Parenthesis are used to indicate a minor component. Two parenthesis indicate a shoulder.

species are different than those observed *in vivo*, perhaps reflecting additional properties of the primary electron-acceptor species [24, 138]. These are discussed in a later section.

The polypeptide components present may reflect the basic structural unit of the phototrap complex, or they may represent, to some extent, contamination with polypeptides that are not required for phototrap function. As may be seen from Table 2, the apparent molecular weight as determined by polyacrylamide gel electrophoresis (PAGE) in solvents containing sodium dodecyl sulfate (SDS) vary according to the particular buffer and procedure used in conducting the electrophoresis (see also Figs. 16 and 18). Because the SDS PAGE method of molecular weight determination depends on complete denaturation and full extension of the polypeptide, it should be applied to analysis of integral membrane proteins with caution—appropriate standards for membrane proteins are not available. The integral membrane proteins have evolved over many years to be compatible with phospholipids in order to carry out their various functions. The likelihood of their resisting full denaturation by SDS seems high. Indeed, there is good experimental evidence that denaturation is incomplete in SDS for many of the peptides

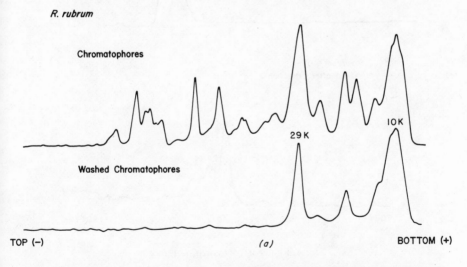

Figure 16. (*a*) Polypeptides observed in chromatophores of *R. rubrum* by SDS PAGE using Tris buffer [95] before and after washing the chromatophores with 0.01 *M* EDTA and 3% Triton X-100. (Taken from [133].) (b) polypeptides observed in photoreceptor complexes (AUT-e) or *R. rubrum* by SDS PAGE using Tris buffer before and after extracting with 1:1 chloroform/methanol. The extracted fraction is also shown. (Taken from [212].)

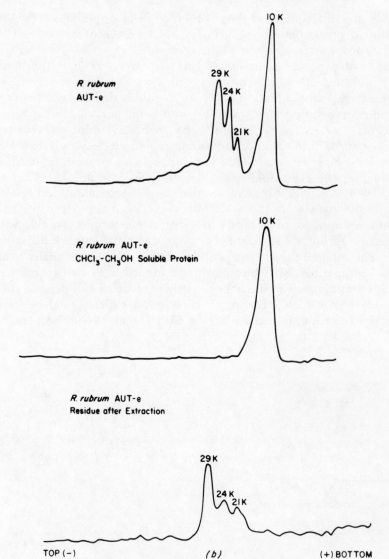

Figure 16.  Continued.

of interest here. I shall return to this problem again in the next section, as well as in the section reviewing reaction center preparations.

Perhaps the ultimate step of isolating each polypeptide and other phototrap componets (e.g., bacteriochlorophyll, bacteriopheophytin, ubiquinone, etc.,), followed by reconstitution of a phototrap indistinguishable from the *in vivo* form will be required before one can fully define the minimal phototrap components. This statement carries with it, of course, the tacit assumption that such a reconstitution is indeed possible. The history of biochemistry suggests that it is inescapable that reconstitution is possible. The question that may be more relevant is one of how long it may take to find the appropriate conditions for reconstitution and to isolate all the needed components with sufficiently high integrity.

*R. rubrum* seems to be one of the simplest known photosynthetic bacteria. Its absorbance spectrum shows one prominent bacteriochlorophyll band in the near infrared, which suggests a single antenna bacteriochlorophyll species. The photosynthetic unit size, which may be defined as the ratio of total bacteriochlorophyll to phototrap, is 28, one of the smallest known (in spinach it is about 300). In accordance with this simplicity, membrane vesicles prepared form this bacteria may simply be washed with 3% Triton X-100 and $10^{-3}$ M EDTA with the result that essentially all polypeptides are removed except those that are thought to play a role in the photoreceptor complex (see Fig. 16). This indicates that the proliferation that occurs in the cell membrane when the bacteria are grown anaerobically under low light intensity results in a very specialized form of membrane in which the photoreceptor complex has a predominent structural and functional role.

*3.1.1.2 Polypeptide Isolation.* Some progress has been made in isolating large quantitites of one of the polypeptides that still remains in purified photoreceptor complexes. During the course of lipid analysis we found [134, 135] that the standard procedures used for lipid extraction (e.g., 1:1 chloroform/methanol) also resulted in extraction of significant amounts of protein. The bacteria had been grown with $^{14}$C-phenylalanine added to their media which is incorporated only into protein. Thus it was easy to assay for protein that might dissolve in organic solvents. Table 3 summarizes the results of extracting dried chromatophores with a variety of organic solvents. One-to-one mixtures of solvents containing methanol were particularly effective. Note that in the best case (1:1 chloroform/methanol) about half the total protein dissolved (according to $^{14}$C-phenylalanine content) in this solvent mixture used to extract membrane fractions from *R. rubrum*. Polypeptides are also obtained with such

## TABLE 3   PROTEIN EXTRACTED FROM CHROMATOPHORES[a]

| Solvent | Percent Total | Solvent | Percent Total |
|---------|---------------|---------|---------------|
| Methanol | 6 | Carbon tetrachloride | 0.3 |
| 2:1 Methanol/chloroform | 46 | Benzene | 0.2 |
| 1:1 Methanol/chloroform | 50 | Acetone | 0.3 |
| 1:2 Methanol/chloroform | 23 | Dimethyl formamide | 2 |
| Chloroform | 0.5 | 1:1 Dioxane/water | 8 |
| 2:1 Methanol/dimethyl sulfoxide | 16 | Dioxane | 0.4 |
| 1:1 Methanol/dimethyl sulfoxide | 25 | 2:1 Methanol/petroleum ether | 16 |
| Dimethyl sulfoxide | 9 | 1:1 Methanol/petroleum ether | 25 |
| Ethanol | 0.8 | Petroleum ether | 0.2 |

[a]Chromatophores were dried by lyophilization and extracted several times with the solvent listed. Soluble and insoluble materials were separated by filtering through a medium sintered-glass filter.

extractions of lyophilized chromatophores from other photosynthetic bacteria, but the percent of total protein dissolving in each case varies widely (e.g., only about 15% of the total protein extracted from chromatophores of *R. spheroides* into 1:1 chloroform/methanol). The protein dissolved in the organic solvent was separated from other components also dissolved in these solvents (e.g., bacteriochlorophyll, carotenoids, ubiquinone, and phospholipids) by column chromatography on Sephadex LH 60, a hydroxypropylated polydextran that can be used for gel filtration in organic solvents.

By using similar extractions and chromatography on purified photoreceptor complexes of *R. rubrum,* together with analytical SDS PAGE, we have been able to show that it is the smaller polypeptide (apparent mol. wt. $\cong$9000-12,000 according to SDS PAGE using Tris buffer) that seems to be totally extracted. Figure 16*b* shows these results. Amino acid analysis demonstrates that the same protein (or polypeptide) dissolves from both intact chromatophores and purified photoreceptor complexes, since the amino acid content is the same whether extracted from either material. According to amino acid analysis (see Table 4) histidine is present in the smallest amount. If it is assumed that there is one histidine/polypeptide, the molecular weight would be about 19,000, which is closer to the value found by SDS PAGE using phosphate buffer. Strong support for this

**TABLE 4   AMINO ACID COMPOSITION OF THE ORGANIC-SOLVENT-SOLUBLE PEPTIDES FROM PHOTORECEPTOR COMPLEXES (MOLE %)**

| | *R. Capsulata* | *R. Rubrum* | *R. spheroides* Inactive Antenna Protein | Active Antenna Protein |
|---|---|---|---|---|
| Lys | 3.9 | 2.0 | 4.4 | 5.7 |
| His | 1.7 | 0.6 | 3.4 | 4.1 |
| Arg | 1.0 | 7.8 | 2.0 | 2.7 |
| Asp | 9.8 | 4.8 | 6.4 | 6.4 |
| Thr | 7.9 | 8.6 | 8.5 | 7.0 |
| Ser | 6.7 | 8.4 | 3.5 | 5.2 |
| Glu | 5.9 | 13.2 | 9.7 | 9.4 |
| Pro | 6.0 | 6.0 | 3.9 | — |
| Gly | 13.6 | 6.3 | 12.0 | 13.4 |
| Ala | 18.8 | 8.7 | 18.6 | 18.6 |
| Val | 2.7 | 4.8 | 6.0 | 4.8 |
| Met | 4.3 | 2.0 | 2.9 | 3.2 |
| Ile | 3.2 | 3.7 | 2.8 | 2.8 |
| Leu | 10.0 | 16.2 | 12.9 | 12.2 |
| Tyr | 1.7 | 0.0 | 0 | 0.7 |
| Phe | 2.5 | 7.3 | 3.0 | 3.8 |
| ½ Cys | 0.0 | 0.0 | 0 | 0 |
| Trp | — | — | — | — |

assignment is provided by the results of CNBr cleavage. From the amino acid analysis, if there is one histidine/polypeptide, then there are three methionines. Since CNBr treatment can be used to cleave quantitiatively a polypeptide at each methionine residue, one would expect to find four shorter polypeptides after such treatment, if only one linear polypeptide chain (containing the three methionines) was initially present. This indeed has been found [134, 135].

The fact that a major polypeptide is selectively extracted with solvents that are routinely thought to remove only pigments and phospholipids from membrane systems should be carefully considered. Many persons who used the SDS PAGE evidence for having achieved a reaction center preparation first extracted phospholipids and pigments from their samples before conducting SDS PAGE on the residue.

An important observation that should be underscored here is the limitation of the SDS PAGE method for molecular weight determination. According to the SDS PAGE method using Tris buffer, the polypeptide isolated above had a molecular weight of about 9000. However, from amino acid analysis of the same component isolated by organic solvent extraction, the minimal molecular weight is 19,000. Because of the results with cyanogen bromide cleavage, the latter molecular weight would appear to be much closer to the actual molecular weight.

Aside from the limitation on using SDS PAGE as a means for molecular weight determination, it is a convenient analytical tool and a given pattern is readily reproduced in different laboratories when exactly the same experimental procedures are followed and when the same material is compared. For example, the three polypeptides originally demonstrated by Feher [95] in purified reaction centers (discussed in more detail later), which were prepared from the R-26 mutant of *R. spheroides* and analyzed by the Tris SDS PAGE method, were verified by Clayton and Haselkorn [96] and have been subsequently reproduced in many other laboratories, including those of this author [133]. Therefore with the recognition that preparations of photoreceptor complexes from *R. rubrum* have given patterns that vary between that shown in Figure 16a, second trace (labeled washed chromatophores), and that of Figure 16b, first trace, this variation must reflect some heterogeneity in the samples from preparation to preparation. Our results with the photoreceptor complex of *R. rubrum* can be made more consistent with respect to the patterns displayed by reaction center preparations if one assumes that (a) the apparently smaller polypeptide (apparent mol wt = 9000-12,000 by Tris SDS PAGE) in photoreceptor complexes is intimately involved in binding the antenna pigments to the phototrap and is either not present in purified reaction center preparations; or is present in modified form (e.g., the apparent mol. wt. = 21,000 polypeptide by tris SDS PAGE); and (b) this smaller polypeptide interacts so strongly with one of the phototrap polypeptides (e.g., the one of mol wt = 24,000 according to Tris SDS PAGE) that the two are not always separated in SDS PAGE and appear as low-molecular-weight components because the complex is not denatured.

### 3.1.1.3 Variability of Proteins in Different Bacteria.

From the point of view of the importance of the phototrap to photosynthesis one might initially think that once the more primative organisms had evolved an effective photosynthetic unit, that unit would of necessity survive in a pretty constant form. Indeed, the physical properties of the phototrap and its pigment composition are very similar over a wide range of photosynthetic bacteria. Therefore one might expect that the protein that specifically binds that phototrap would also appear in very similar form in all species.

In this regard it may be useful to examine, for example, the constancy of a polypeptide chain length and amino acid composition for a well known protein that is present in the photosynthetic bacteria of interest here and whose function seems to be the same in all these organisms. Fortunately, extensive information is available [142] for such a protein, which, in fact, seems to play the role as the secondary electron donor in these photosynthetic organisms, namely, cytochrome c (cytochrome $c_2$ in *R. rubrum*). This electron transport component is uniformly found in all known electron transport systems involving oxidative or photophosphorylation. Its absorbance spectra, the fifth and sixth ligands coordinated to iron, the thioether linkages of the iron porphyrin to the polypeptide backbone, the acid-base stability, heat stability, and many other related properties are extremely uniform across the biological world. In spite of this similarity of function and physical properties, only 30% of the amino acid residues are homologous out of nearly 80 cytochromes of the c class that have been sequenced [143]. In addition, the chain length of the polypeptide varies from 8500 to 50,000, although the variation within the photosynthetic bacteria may only be from 10,000 to 13,000.

It is also well known that enzymes that apparently catalyze the same reaction and exhibit high conservation of active site components and mechanism of catalysis nevertheless show great variation in polypeptide chain length.

Based on the above observations, it may not be unexpected that the polypeptides responsible for holding phototrap pigments in a particular configuration in different photosynthetic bacteria or their mutants may have substantial differences in amino acid content and polypeptide length. A further point that should perhaps be made is that although two photosynthetic bacteria such as *R. rubrum* and *R. spheroides* may appear superfically to be closely related, the time since they diverged from a common ancestor may be of the order of a billion years.

Significant difference among the photosynthetic bacteria are seen in the relative amount and amino acid content of the polypeptide in photoreceptor preparations that are soluble in organic solvents. Table 4 compares the amino acid content of several such preparations from different bacteria. Not only are there large differences in the relative amino acid content, but, interestingly, the polarity is significantly

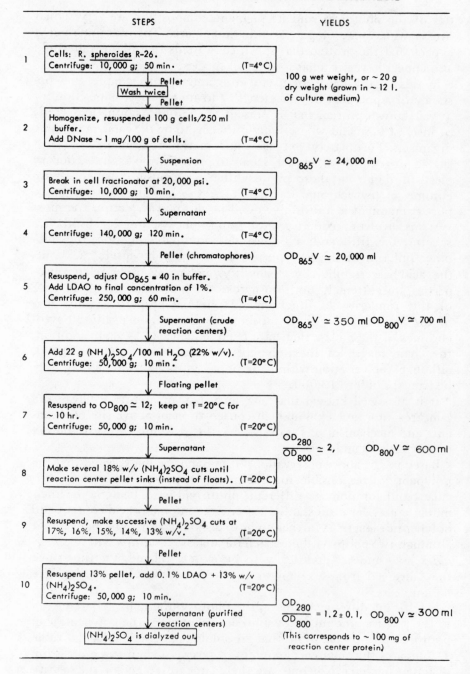

STEPS                                                                    YIELDS

1   Cells: R. spheroides R-26.
    Centrifuge: 10,000 g; 50 min.                    (T=4°C)
                                                                         100 g wet weight, or ~ 20 g
                              Pellet                                     dry weight (grown in ~ 12 l.
                              [Wash twice]                               of culture medium)
                              Pellet

2   Homogenize, resuspended 100 g cells/250 ml
      buffer.
    Add DNase ~ 1 mg/100 g of cells.                 (T=4°C)

                              Suspension                                 $OD_{865}V \simeq 24,000$ ml

3   Break in cell fractionator at 20,000 psi.
    Centrifuge: 10,000 g; 10 min.                    (T=4°C)

                              Supernatant

4   Centrifuge: 140,000 g; 120 min.                  (T=4°C)

                              Pellet (chromatophores)                    $OD_{865}V \simeq 20,000$ ml

5   Resuspend, adjust $OD_{865}$ = 40 in buffer.
    Add LDAO to final concentration of 1%.
    Centrifuge: 250,000 g; 60 min.                   (T=4°C)

                              Supernatant (crude                         $OD_{865}V \simeq 350$ ml $OD_{800}V \simeq 700$ ml
                              reaction centers)

6   Add 22 g $(NH_4)_2SO_4$/100 ml $H_2O$ (22% w/v).
    Centrifuge: 50,000 g; 10 min.                    (T=20°C)

                              Floating pellet

7   Resuspend to $OD_{800} \simeq 12$; keep at T=20°C for
    ~ 10 hr.
    Centrifuge: 50,000 g; 10 min.                    (T=20°C)

                              Supernatant                                $\dfrac{OD_{280}}{OD_{800}} \simeq 2,$     $OD_{800}V \simeq 600$ ml

8   Make several 18% w/v $(NH_4)_2SO_4$ cuts until
    reaction center pellet sinks (instead of floats). (T=20°C)

                              Pellet

9   Resuspend, make successive $(NH_4)_2SO_4$ cuts at
    17%, 16%, 15%, 14%, 13% w/v.                     (T=20°C)

                              Pellet

10  Resuspend 13% pellet, add 0.1% LDAO + 13% w/v
    $(NH_4)_2SO_4$.                                   (T=20°C)
    Centrifuge: 50,000 g; 10 min.

                              Supernatant (purified                      $\dfrac{OD_{280}}{OD_{800}} = 1.2 \pm 0.1,$  $OD_{800}V \simeq 300$ ml
                              reaction centers)
                              [$(NH_4)_2SO_4$ is dialyzed out.]           (This corresponds to ~ 100 mg of
                                                                         reaction center protein.)

different even though they are all soluble in organic solvents.

Variations in the apparent size and number of polypeptides in photoreceptor preparations from different bacteria are more extensively compared after reaction center preparations are discussed in the next section.

### 3.1.2  Reaction Center Preparations.    3.1.2.1  Preparation and Analysis.

The first efforts to isolate a reaction center yielded a preparation that may best be described as somewhere between an unmasked phototrap and a purified reaction center [93]. Later refinements and additional purification procedures [94-96, 132] provided the first well characterized reaction center preparations.

A schematic outline of a typical reaction center preparation is shown in Figure 17. The type of bacteria with which this kind of phototrap preparation has been best documented are carotenoidless mutants of *R. spheroides* or *R. rubrum* [94-96, 132]. Recent reports indicate that similar preparations can also be achieved using wild type bacteria [144-150, 198, 245, 246]. In some of the latter cases further documentation is desirable to distinguish between true isolation of a reaction center and obtaining an unmasked phototrap; both the antenna pigment solubilization and/or degradation and the polypeptide-organic solvent solubilization problems alluded to earlier need to be better documented with regard to these preparations.

Properties of the reaction center preparation from the R-26 mutant of *R. spheroides* have been well characterized [94-96, 132]. *In vivo* light-induced absorbance and esr changes were demonstrated to occur with a quantum yield near 1.0 [88, 89]. Low-temperature formation and decay kinetics of the photoproduced esr signal were especially carefully studied and shown to be the same as in the *in vivo* system [79, 80]. Analytical data are summarized in Table 5. No significant phospholipid, carbohydrate, or transition metals are present, except for iron which can be removed along with one polypeptide by further purification [132, 151]. The redox properties of the primary electron-donor unit were not determined.

At the preparative stage where there was one iron per phototrap,

---

Figure 17.  Flow chart of the steps used to obtain purified reaction centers from *R. spheroides* R-26. Buffer: $0 \cdot 01$ $M$ Tris-Hcl (pH 8), with $0 \cdot 5$ m$M$ EDTA and and $0 \cdot 5$ m$M$ EGTA added for steps 7-10. OD$_{865}$ $V$ is the optical density at 865 nm in a 1-cm path length, times the volume of the solution; this is a convenient unit since it is independent of dilution and reflects the total amount of material at hand [95].

## TABLE 5   ANALYSIS OF REACTION CENTER COMPLEXES
### FROM *R. SPHEROIDES* R-26 (95, 96, 132)

| | | | |
|---|---|---|---|
| Molecular weight[b] | 80,000 [95] | | 51,000 [132] |
| Bacteriochlorophyll $a^a$ | 4 | [94-96, 99] | 4 |
| Bacteriopheophytin $a^a$ | 2 | [94-96, 99] | 2 |
| Carotenoid[a] | 0 | | 0 |
| P-lipid[a] | — | | — |
| Ubiquinone[a] | 1 | | — |
| Iron[a] | 0.8 | | 0.05 |
| Copper[a] | <0.2 | | — |
| Manganese[a] | 0.2 | | — |
| φ for P865 oxidation | 0.9 | [88] | — |
| $E'_0$ for P865 | — | | — |
| Labile sulfide[a] | <0.01 | | — |
| Carbohydrate | <0.3% (w/w) | | — |
| Polypeptides | 28.000[c] | | 24,000[c] |
| | 24,000[c] | | 21,000[c] |
| | 21,000[c] | | |

[a] Estimated number per phototrap.
[b] Estimated by adding one of each polypeptide and the pigments indicated.
[c] Tris method of SDS PAGE [95, 96, 132].

there were also three polypeptides whose 1:1:1 stoichiometry has been carefully documented [98, 132]. In addition, Steiner et al. [98] have been able to isolate sufficient quantities of all three polypeptides to establish the amino acid content of each. These are reproduced in Table 6. The amino acid compositions of the smaller two polypeptides are quite similar and show a very low polarity. The smallest well characterized active phototrap preparation [132] contains only these two polypeptides. The minimal molecular weight of this phototrap would be expected to be at least 21,000 + 24,000 + 4000 (for the donor unit bacteriochlorophyll) + 2000 (for the two bacteriopheophytin present), or about 51,000. In this connection it is interesting to note that the active phototrap preparation containing only the two polypeptides can be observed still intact and active on the polyacrylamide gel when low SDS concentrations are used (e.g., .05%) or when normal SDS concentrations are used in the presence of a nonionic detergent. The nonionic detergents [132] seem to have an antagonistic effect on the denaturation caused by SDS. Under such conditions the polypeptide appears as a band of mol wt = 37,000, some 14,000 daltons lower than that indicated above for the minimal size. Together with the earlier example of the

**TABLE 6  AMINO ACID COMPOSITION OF REACTION CENTER POLYPEPTIDES FROM** *R. SPHEROIDES* **R-26 [98]**

| Amino Acid | Polypeptide Molecular Weight | | |
|---|---|---|---|
| | 21,000 | 24,000 | 28,000 |
| | (mole %) | (mole %) | (mole %) |
| Lys | 2.4 | 0.9 | 5.1 |
| His | 2.3 | 2.1 | 2.3 |
| Arg | 3.3 | 4.0 | 4.4 |
| Asp | 6.3 | 6.4 | 8.2 |
| Thr | 5.2 | 4.4 | 4.7 |
| Ser | 4.4 | 5.6 | 5.0 |
| Glu | 4.8 | 6.5 | 8.2 |
| Pro | 5.9 | 4.7 | 7.8 |
| Gly | 12.0 | 11.7 | 10.3 |
| Ala | 9.7 | 9.9 | 10.5 |
| Val | 6.0 | 5.5 | 7.8 |
| Met | 1.6 | 3.1 | 2.0 |
| Ile | 6.2 | 5.1 | 5.4 |
| Leu | 11.2 | 11.9 | 9.9 |
| Tyr | 4.3 | 3.3 | 2.5 |
| Phe | 7.5 | 7.7 | 4.2 |
| ½ Cys | 1.3 | 0.0 | 0.9 |
| Trp | 5.7 | 7.3 | 0.8 |

behavior of the organic solvent-soluble polypeptides from photo-receptor preparations on SDS PAGE, this above result again points out the danger in assuming that the SDS PAGE method is entirely valid for determining molecular weights of membrane proteins. In these two particular examples, the size estimated by SDS PAGE was probably in error by at least 30% and may have been low by a factor of 2.

Several buffer systems have been used in SDS PAGE in which photosynthetic membranes have been studied. Feher et al. [95, 132] and Clayton and Haselkorn [96] used tris-(hydroxymethyl)-amino-methane as the buffer, while others have used the Weber and Osborn procedure [134, 144, 153] in which the buffer is phosphate. In our work results in the two buffers have been substantially different (see Figs. 16 and 18). Although it is not proven in which buffer system the polypeptides are more denatured, and therefore would yield more valid molecular weights, our results with the Weber and Osborn method gave a molecular weight for the organic solvent soluble

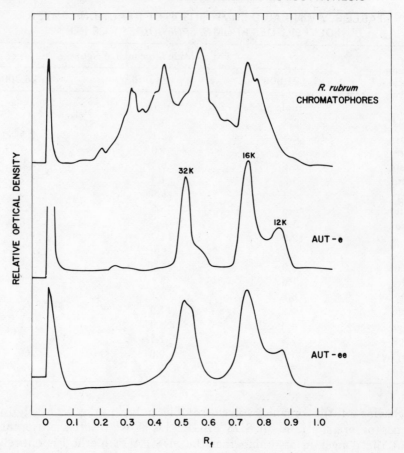

Figure 18.   Polypeptides observed in chromatophores and photoreceptor complexes from *R. rubrum* as determined by SDS PAGE using a slightly modified Weber and Osborn method [141, 133].

peptide closer to that of the isolated polypeptide. Tris buffer is a polyol that has deleterious effects on membrane-related phenomena at higher concentrations (e.g., photophosphorylation and oxygen evolution). The conjugate base form may be active as a nonionic detergent, which in SDS PAGE antagonizes the denaturing effect of SDS, similar to Triton X-100.

If the current SDS PAGE results reflect proteins that are incompletely denatured, the possibility exists that multiple bands may reflect the same peptide but in different states of denaturation. This does not seem to be the case for the reaction center preparation, since, as Steiner et al. [98] point out, the larger appearing polypeptide (mol wt = 24,000 by Tris SDS PAGE) contains only half as much lysine as the smaller appearing polypeptide (mol wt = 21,000 by Tris SDS PAGE), and there was no cysteine in the larger polypeptide but two in the smaller one. However, it is probably not yet valid to assume that the polypeptide characterized as 21,000 mol wt by Tris SDS PAGE is indeed smaller than the one characterized as mol wt = 24,000 by the same method. As may be seen in Table 6, their amino acid composition on a weight basis is highly similar.

### 3.1.2.2 Physiological Role of the Organic-Solvent-Soluble Polypeptides in R. spheroides.

Application of the AUT procedure to chromatophores of *R. spheroides* followed by column electrophoresis yields a photoreceptor complex in which the antenna pigments are loosely bound to the phototrap [133]. Interestingly, the polypeptides observed by SDS PAGE had apparent molecular weights of 24,000, 21,000, 12,000, and 9000. Note that this is a somewhat different pattern than is observed with the photoreceptor preparation from *R. rubrum* (Fig. 16). At the very least, there was no apparent 30,000 molecular weight polypeptide in the photoreceptor preparation from *R. spheroides*. In addition, a reaction center having only the 24,000 and 21,000 polypeptides (similar to that obtained in reaction center preparations cited in Table 6) partially separated during electrophoresis, leaving an antenna-rich fraction containing primarily the 9000 and 12,000 polypeptides. Since the phototrap seems to function appropriately without the latter polypeptides, there is once again the implication that these small polypeptides are responsible for binding the antenna bacteriochlorophyll aggregate and carotenoid components and also interact specifically with the reaction center.

### 3.1.2.3 Antigenic Uniqueness of Reaction Center Protein.

Further indication that significant differences exist in the protein of reaction centers isolated from different bacteria has been provided by immunological analyses. Clayton and Haselkorn [96] and Clayton and Clayton [154] prepared antisera to their reaction center preparation from the R-26 mutant of *R. spheroides* and showed that it would not cross-react with preparations from other photosynthetic bacteria, even *R. capsulata*. It would react, however, with detergent-treated chromatophores of wild type *R. spheroides*, causing a precipitate containing the phototrap and leaving antenna pigments in the supernatant. Steiner et al. [98] showed that antisera to reaction centers prepared from the G-9 carotenoidless mutant of *R. rubrum* and the R-26 carotenoidless mutant of *R. spheroides* formed precipitates in the immunodiffusion assay only with the homologous antigens.

Since it is known that antibodies are formed against several different hapten groups, even on a small protein like cytochrome c [155], it seems likely that because of the lack of cross-reaction to antiserum in the above experiments, conducted in two independent laboratories, there are major differences in amino acid sequences and/or three-dimensional structure in the polypeptides of the phototrap as found in different bacteria.

### 3.1.3. Evolutionary Pattern of Bacterial Photosynthesis.
Almost from the time photosynthetic bacteria were first isolated, it was believed that those with one prominent bacteriochlorophyll band (e.g., *R. rubrum*) in the near infrared were more primitive than those with two (e.g., *R. spheroides*) or more (e.g., *Chromatium*) major bands. Consistent with this idea is the fact that electron transport components also seem to become more complicated as one examines *R. rubrum, R. spheroides,* and *Chromatium* in that order. Again, there is a progression in the photosynthetic unit size in these three bacteria from about 28 to 50 to 80, respectively.

Of interest from this point of view are the observations that a single photoreceptor complex of small photosynthetic unit size was prepared from *R. rubrum* by the AUT-e method, whereas two distinct bacteriochlorophyll-protein complexes were prepared from *R. spheroides* by the AUT-e method [133]. Only one of the latter two had phototrap activity, and the polypeptide content of each appeared to differ according to SDS PAGE. The ratio of bacteriochlorophyll to phototrap in the active fraction was about 30. Thus one might conclude that there may be a fundamental photoreceptor complex in all photosynthetic bacteria that has a photosynthetic unit size like that of *R. rubrum,* about 28 BChl per trap. The additional inactive antenna-protein complex found in *R. spheroides* may represent an extended antenna, which would be an advantage adopted by these cells as they evolved. Thus, with more light-harvesting pigments able to funnel their energy into one phototrap, the cell could save on the number of required phototraps as well as associated electron transport and phosphorylation components. In this regard, it would be interesting to apply the AUT-e method to the seemingly even more complicated *Chromatium.*

It has been reported that the ratio of bacteriochlorophll to phototrap increases significantly in *R. spheroides* but not in *R. rubrum* if the cells are grown in dim light as compared with bright light [156]. This would suggest that the innactive antenna-protein complex might be under genetic control separate from that of the photoreceptor complex and thus afford a more efficient way for the cell to adapt to its lighting conditions. Unfortunately, in spite of several attempts we have not been able to reproduce this experiment with our strain of *R. spheroides* under our growing conditions. If such a variation does occur under appropriate conditions, it should be interesting to compare the relative amounts of active and innactive antenna-protein complexes as obtained by the AUT-e method from cells grown under dim and bright light.

There are several indications that the two antenna-protein complexes of *R. spheroides* interact in the membrane. One of these comes from quantum yield measurements [137] for the photooxidation of the primary electron-donor unit. The quantum yield for intact chromatophores was 0.9, while that for the same sample after AUT treatment was 0.5. Presumably, in dilute solution the innactive antenna is separated too far from the photoreceptor complex to pass its excitation energy on to a phototrap-containing complex. Reaggregation by removal of detergent results in an increase once again in the quantum yield.

A second property of these bacteria that might be used to indicate the interaction between the inactive and active antenna-protein complexes is an absorbance band at 890 nm that appears as a prominent shoulder in intact cells and freshly prepared chromatophores. Treatment with low concentrations of Triton X-100 causes this band to disappear and the ratio of the 850 to 800 nm bands changes slightly [137]. This seems to be a reversible effect, since the 890 nm band is restored on removal of the detergent. Rather than attributing the 890 nm band to a unique bacteriochlorophyll-protein complex, as has been suggested by some, I would like to suggest that it arises because of an extension of the size of a hydrated bacteriochlorophyll aggregate or due to bacteriochlorophyll-bacteriochlorophyll interaction on adjacent protein-pigment complexes. This could occur as the two antenna-protein complexes (one with an active phototrap and one without) interact with each other. The connecting bacteriochlorophyll molecules would be highly sensitive to detergents and perhaps to any three-dimensional changes in the protein that must occur when the phototrap is excited [24].

Following from the last comment is an additional suggestion as to the source of perhaps some of the bacteriochlorophyll band shifts and some of the carotenoid band shifts that occur when the phototrap is excited, or when an ion gradient is imposed on the membrane [28]. Rather than explaining a band shift as a result of a field effect due to charge spearation in the vicinity of the phototrap, or across the membrane, one might consider that the interaction between the two antenna-protein complexes changes because of a three-dimensional change in the conformation of the phototrap, thus disrupting the extended bacteriochlorophyll interaction (and possibly carotenoid-carotenoid interaction).

### 3.1.4 Other Polypeptide Isolation from Photosynthetic Bacteria.

The photosynthetic bacteria have been of interest to microbiologists and bacterial physiologists because of the uniquely different membrane structure and metabolism depending on the growing conditions of these microorganisms. The purple bacteria, such as *R. rubrum* and *R. spheroides*, can grow anaerobically in the presence of light (phototrophic) or aerobically in the dark (chemotropic). Only under conditions of anaerobic growth does the cell membrane proliferate into the cytosol of the cell, as seen in Figure 3. The transition of such bacteria from chemotropic to phototropic growth offers the opportunity to study the induction and development of the complex cell membrane. As a result many studies by microbiologists [156-182] of the structure and development of these membranes have paralleled the previously reviewed studies on the isolation of the photochemical machinery.

Fraker and Kaplan [174, 175] have developed a method for fractionating chromatophore protein by dissolving membrane vesicles of *R. spheroides* in 2-chloroethanol containing approximately 1 $M$ HCl. Clear solutions were obtained, but of course phototrap activity was irreversibly lost; phototrap activity is lost either by treatment with such an organic solvent, or at pH values below 4. In addition, bacteriochlorophyll is converted to bacteriopheophytin under such acidic conditions. Soluble protein was separated from insoluble protein by centrifugation at 27,000$g$ for 15 min. Several major polypeptides existed in this fraction according to SDS PAGE and one of these seemed to have a pigment complement associated with it. This polypeptide was isolated by preparative SDS PAGE and found to contain 59% protein, 35% phospholipid, and 6% bacteriopheopytin. SDS PAGE gave an apparent molecular weight of 10,000. In view of the foregoing discussion regarding denaturation of membrane proteins by SDS and the existence of two different antenna pigment-protein complexes in *R. spheroides*, it would be very surprising indeed if the above protein fraction contained a single component; even the phospholipids and pigments are still part of the complex. Also, the amino acid content is very similar to that of intact chromatophores. Nevertheless, this fraction has been further studied to determine a C-terminal amino acid sequence and to cleave at methionine residues by treatment with cyanogen bromide [176-178]. The authors estimate that this fraction represents more than 50% of the total membrane protein. These results are difficult to reconcile with those previously outlined [133] for AUT-e treatment of the same wild type strain of R. *spheroides*. In the latter work two antenna-protein complexes were isolated (one with an active photo-trap and one without), both of which had several apparently different polypeptide components. Hopefully, further work will clarify the relationship between these preparations.

Another antenna-protein complex has been isolated from R. *spheroides* by Clayton and Clayton under conditions used for reaction center preparations [154]. One procedure, which resulted in a product with the fewest polypeptides, involved the utilization of LDAO (lauryl dimethyl amine oxide) and $(NH_4)_2SO_4$ to collect an insoluble floating protein-pigment complex. This step was followed by dispersion in Triton X-100 and centrifugation to separate an antenna complex (pellet) from reaction centers (super-natant). The pellet contained bacteriochlorophyll and carotenoid and had an absorbance spectrum similar to the *in vivo* antenna complex. Analysis of the protein by the Tris buffer method of SDS PAGE showed a major polypeptide of an apparent molecular weight of 9000 which still retained most of the pigments. This latter fact suggests that the complex was not truly denatured under conditions of SDS PAGE. With wild type R. *spheroides* about 60% of the original light harvesting bacteriochlorophyll was contained in this prepara-tion. Comparison of these results with the earlier discussed AUT-e treatment suggests that the protein fraction prepared by Clayton and Clayton [154] may be the same as the inactive pigment-protein complex isolated by the AUT-e method.

Yet a different pigment-protein isolation procedure was recently reported that results in preparation of a carotenoprotein from R. *rubrum*. Schwenker et al. [183] used either an ammoniacal acetone extraction of chromatophores or SDS solubilization. By the latter method they dissolved the chromatophores in 1% SDS containing 1% β-mercaptoethanol and centrifuged the solution. The supernatant was fractionated with $(NH_4)_2SO_4$ and the carotenoprotein fraction was subjected to gel filtration chromatography (Sepharose 6B) in a buffer containing 0.01% SDS. SDS PAGE analysis gave a single band of apparent molecular weight = 11,000. One carotenoid pigment molecule was found bound to the protein (mol wt = 11,000), and it was indicated to be spirilloxanthin according to its behavior on thin layer chromatography. Appreciable carbohydrate was also present and it was estimated that this protein represented at least 10% of the total protein in the membrane. Schwenker et al. [183] did not feel they could rule out the possibility that the preparation was artifactual. Interestingly, if all the carotenoids in the membrane of this bacteria were complexed in such a 1:1 fashion, more than 25% of the total protein content of the membrane would have to be carotenoprotein, since there are about 10 carotenoid molecules per phototrap. Since this does not seem likely, the carotenoprotein would rather seem to represent a remnant of a more extensive pigment complex—it may not be artifactual per se, but rather it may have retained only part of its original pigment content, and perhaps only part of the polypeptide complement of the original protein.

### 3.1.5 Photoreceptor Complexes and Membrane Structure.

An apparently unique feature of the photosynthetic membrane of *R. rubrum* is that one polypeptide, the one that is soluble in organic solvent, accounts for about 50% of the total membrane protein by weight. Approximately another 25% of the total protein would seem to be accounted for by the other polypeptide isolated in the photo-receptor complex, that is the component of mol wt = 32,000. Thus complete characterization of the photoreceptor complex of *R. rubrum* will also result in a knowledge of much of the structure of a specialized membrane system.

Although any attempt to suggest the three-dimensional structure of the photoreceptor complex and its relationship to the membrane in which it is found must be almost entirely speculative, a working model does serve the purpose of summarizing important features of the system and also may suggest new experiments. Figure 19 shows top and side views of a photoreceptor complex. The figure portrays a "duplex system" for reasons explained later in the section dealing with the mechanism of the primary photochemical event. For the present purposes it is sufficient to consider one antenna complex and one donor unit complex.

The antenna complex portrayed in Figure 19 is arbitrarily shown to extend across the membrane involving four polypeptides of mol wt = 19,000. Recent circular dichroism measurements in our laboratory [135] indicate that this peptide contains an extensive α-helical content in SDS or organic solvents. A distinct feature of the model is that a single polypeptide does not necessarily bind a specific number of bacteriochlorophyll molecules. Rather, several polypeptides acting in concert may provide a "cage" for the pigment complex. While the figure does not portray a single polypeptide extending entirely across the membrane, at this juncture that possibility would also seem to be a suitable alternative. Carotenoid molecules are utilized to stabilize the entire complex because their length, rigidity, and nonpolar nature seem idealy suited to interact with the α-helix portion of the polypeptide having hydrophobic side chains. The bacteriochlorophyll molecules are assumed to occur in two aggregates of 12 molecules each, separated by the hydrophobic center of the membrane to which the geranylgeranyl tails [77] of the bacteriochlorophyll also contribute. If one feels uncomfortable about the complex spanning the membrane, the two aggregates of 12 antenna bacteriochlorophyll molecules could also be envisioned to occur on the same side of the membrane with the aggregate spreading in two dimensions (this could be best envisioned by intermeshing a second aggregate of 12 with that already shown in the top view of Fig. 19). That part of the polypeptide facing the aqueous phase would have many polar amino acid side shains, while those toward the center of the membrane, and in the helical region, would contain only nonpolar amino acid side chains. An amino acid sequence of this important polypeptide would certainly provide a good test of the model. Consistent with the model is the fact that carotenoidless mutants have much more labile antenna complexes than their wild type counterparts and the antenna is also much more readily separated from the phototrap complex.

The phototrap complex is suggested to bind to the antenna complex by specific interaction of the polypeptide portions of each complex, possibly facilitated by carotenoid or bacteriochlorophyll molecules. A single polypeptide of apparent mol wt = 32,000 is arbitrarily used in the figure for housing the phototrap components. For reasons discussed later, the primary electron-acceptor molecule (ubiquinone) is indicated as being shared by two primary electron-donor units. The phototrap bacteriochlorophyll molecules are shown to be near the surface of the membrane where they may undergo exchange [24] or, in the *in vivo* situation, they may interact with cytochrome $c_2$ in a secondary electron-transport reaction. The ubiquinone molecule is situated toward the middle of the membrane, perhaps interacting with a bacteriopheophytin molecule to facilitate an electron "cascade", as discussed later. Once reduced, the ubiquinone, if protonated and neutral, may be able to "swing" to a secondary electron acceptor and/or to the other surface of the membrane, thus accomplishing the first step of a proton pump [228].

The model shown in Figure 19 could be enlarged to include a second antenna complex (such as is found in *R. spheroides*) which

TOP VIEW

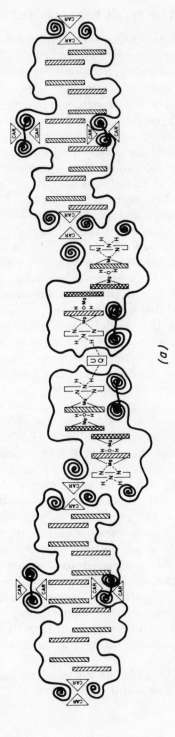

(a)

Figure 19. Hypothetical relationship of the photoreceptor complex to the membrane. The picture portrays a surface view from outside (a) and a crosssectional view of the membrane (b). The membrane structure is assumed to consist of a lipid bilayer interupted by protein-pigment domains, such as those projected here. The thicker lines were used to represent the polypeptide backbone of the protein and the coiled portion to represent α-helical parts of the protein. The bacteriochlorophyll molecules are represented as rectangles with spacing suggestive of a molecular aggregate such as shown in Figure 7 and 8. Bacteriochlorophylls of the antenna and two of those in the phototrap are shown with diagonal lines. The other two bacteriochlorophyll molecules in the phototrap are shown with cross-hatching. The two bacteriopheophytins of the phototrap are not shaded but have three nitrogens of the pyrolle rings represented, the fourth being assumed to be behind the one in the middle. The two hydrogens shown on the bacteriophenophytin nitrogens are therefore to indicate one hydrogen on each of two pyrolle nitrogens, one behind the other. Carotenoid molecules are drawn in close interaction with the nonpolar α-helix regions of the protein with an assumed intercalation of the methyl groups and the amino acid side chains. Additional description of the figure is given in the text.

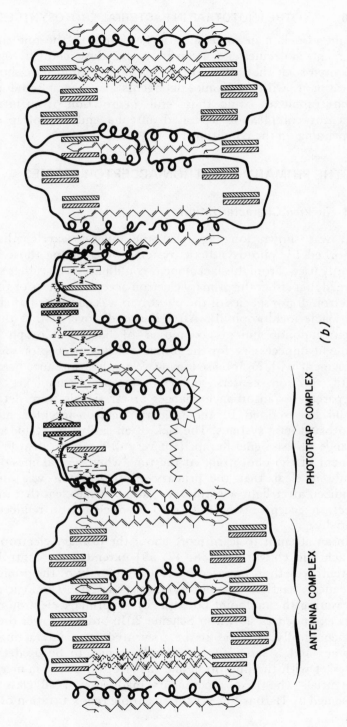

SIDE VIEW (THROUGH MEMBRANE)

PHOTOTRAP COMPLEX

ANTENNA COMPLEX

(b)

147

might associate in some specific fashion with the one shown. The pigment molecules are assumed to be close enough together and oriented such that efficient transfer of an excited state should occur. Treatment with nonionic detergents of a membrane containing such complexes could thus result in isolation of an intact photoreceptor complex or separated antenna and phototrap complexes depending on the conditions.

## 4   THE PRIMARY ELECTRON-ACCEPTOR SPECIES

### 4.1   *In Vivo* Characterization

It was shown long ago that the longest wavelengths of light absorbed by photosynthetic systems were among those most efficiently used. From this fact alone it could have been rather safely concluded that either the primary electron-acceptor species or the primary electron-donor species of the phototrap was composed of chlorophyll (or bacteriochlorophyll). After it was discovered that the primary electron-donor unit is composed of bacteriochlorophyll, it was thought unnecessary for the primary electron-acceptor species to be a pigment with an intense long-wavelength absorbance band. Indeed, until reaction centers and photoreceptor complexes had been prepared, no absorbance changes or esr signals were detected that could be ascribed to the primary electron-acceptor molecule in photosynthetic systems. The lack of an easily observable absorbance change or esr signal has made it very difficult to identify this component of the phototrap. At the time when several lines of evidence made it clear that the primary electron donor was one-electron oxidized after light absorption, it also became clear that the primary electron acceptor would have to be one-electron reduced, at least initially.

In an attempt to learn more about the primary electron acceptor, Loach and co-workers [38, 47, 55] extended the controlled redox potential method initiated by Kok to a study of the primary photochemical reactions in photosynthetic bacteria. The logic applied to measuring the midpoint potential of the primary electron-donor unit was extended as shown in Scheme 2. In order to lower the potential systematically, air had to be prevented from interacting with the system and a series of redox buffers had to be used that would interact with the phototrap and the electrodes used in measuring the potential. The first experiments made use of all glass apparatus designed by Harbury [234] for anaerobic redox titration of oxidative

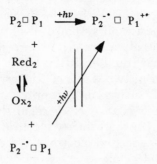

Scheme 2

enzymes. For application of these anaerobic techniques to measurements of the small absorbance changes in photosynthesis it was necessary to construct an instrument [40] that could make quantitative measurements of absorbance changes as small as $10^{-4}$. The anaerobic apparatus was also adopted for taking esr mearurements at known potentials [47]. By monitoring one of the light-induced changes in the primary electron-donor unit, a reversible loss in phototrap activity was found to occur with a midpoint potential of about $-0.01$ V at pH 7.5 (38, 47, 55). An example of data obtained in this kind of experiment is shown in Figure 20.

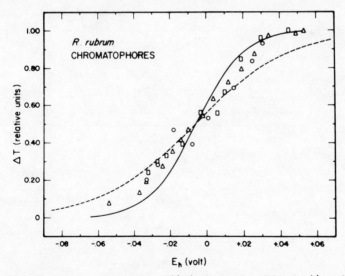

Figure 20.  Changes in transmittance ($\Delta T$) observed at 761 nm ($\Delta$ and $\bigcirc$) and 790 nm ($\square$) during a low-potential oxidation-reduction titration of *R. rubrum* chromatophores at room light. ($-$) Theoretical two-electron reactions; ($---$) theoretical one-electron reaction. (Taken from [47].)

Although it was recognized that the primary electron acceptor in bacterial photosynthesis and system I of oxygen-evolving organisms would have to be initially a one-electron reduced species, no radical had been observed that could be attributed to this species. This could be explained in either of two ways: (1) the primary electron acceptor is a transition metal, such as iron, whose associated esr signal is so broad and absorbance change is so small that they had not been detected [25,47] or (2) the primary electron acceptor is an organic radical that picks up a second electron very rapidly, or very rapidly passes its unpaired electron to a secondary component, such as a transition metal center [25, 47].

## 4.2   Isolated Phototrap Complexes

*4.2.1* **An Iron esr Signal at Low Temperature.** When it became possible to isolate fully active reaction center and photoreceptor complexes, several different laboratories reported the observation of one-electron-reduced species in these systems. Using a highly purified reaction center preparation and applying esr spectroscopy, McElroy et al. [184] reported a new esr signal observed at 1.4°K whose spectrum is shown in Figure 21. They used a uniquely modified esr

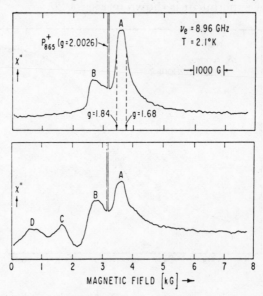

Figure 21. Light-induced esr spectrum of an electron acceptor observed in bacterial-reaction centers at 1.4°K by Feher [95,185]. Reaction centers were prepared from bacteria grown in normal (top) and low iron (bottom) media.

spectrometer to detect this signal. In this spectrometer the normal 100 kHz field modulation was replace by a 10 Hz light-modulation system with superheterodyne detection. With this system only paramagnetic signals produced in reversible photochemical reactions are detected. The replacement of field modulation with light modulation means that absorption of the radical and not its first derivative is detected. Use of a spectrometer that records absorption is very advantageous for both detection of broad paramagnetic signals and for concentration determinations.

At X-band ($\nu$ = 9 GHz) three absorption peaks were detected at magnetic field values of 1.8, 2.9, and 3.7 kG, corresponding to approximate $g$-values of 3.6, 2.2, and 1.8, respectively. Recording of the signal at Q-band ($\nu$ = 35 GHz) showed that only the two highest field X-band peaks shifted with magnetic field from 2.9 and 3.7 kG to 12.1 and 13.5 kG [the intense absorption in the center of the spectrum ($g$-factor = 2.0026) of Figure 21 is due to the bacterio-chlorophyll cation of the donor unit]. The broadness and position ($g$-value) of the esr signal and metal analysis data showing that iron was the only transition metal present with a concentration equal to the reaction center aided Feher in the conclusion that the broad signal was due to a reduced nonheme iron protein. Integration of the areas under the absorption curves for both the cation donor and the reduced broad iron-protein signal yielded approximately a 1:1 concentration ratio.

In a recent extension of these measurements, Feher et al. [185] have been able to substitute Mn(II) in place of part of the iron by growing the bacteria in low-iron media. Using this technique, they have demonstrated that the light-induced signals at 0.7 and 1.8 kG (Fig. 21) were due to manganese in the sample. Furthermore, because the kinetics of decay of the light-induced signals assigned to either iron or manganese did not depend on the extent of manganese substitution, Feher et al. concluded that iron could not be the primary electron acceptor, since a significant change in kinetics of decay would be expected if the electron returned to [BChl]$_{2-4}^{+\bullet}$ directly from the metal.

Investigation of low-temperature esr signals has also been undertaken by Leigh and Dutton [113]. Using the technique of flash photolysis esr spectroscopy [78], they have observed a coincidence in the rate of decay of a signal with a $g$-factor = 1.82 (interpreted by them to be an iron-sulfur protein) and a rate of decay of [BChl]$_{2-4}^{+\bullet}$. The appearance of the $g$ = 1.82 signal in the dark when the redox potential was lowered matched the quenching of phototrap activity and also followed the same titration curve for the formation of the triplet state (see Fig. 14) of the donor unit. Considering all their data, Dutton and Leigh [186] concluded that the $g = 1.82$ signal is the primary electron acceptor and have called it photoredoxin. It is generally agreed that the $g = 1.82$ signals observed by Leigh and Dutton correspond to the original high field absorption observed by Feher et al. at 3.7 kG ($g$-factor = 1.8).

As mentioned earlier, Feher et al. have provided recent evidence that the $g = 1.82$ signal in reaction centers prepared from the R-26 mutant of *R. spheroides* is not consistent with an iron-sulfur protein playing a role as the primary electron-acceptor species. Other difficulties with such an interpretation may also be cited. The fact that the above data, implicating an iron-sulfur protein as the primary electron acceptor species, can only be obtained at temperatures below $20°K$ is somewhat troublesome. This is particularly true when one considers that the crucial room-temperature measurements of characteristic absorbance changes in the 350 to 500 nm region, which would be required for an iron-sulfur protein [187-189], have not been observed even under appropriately redox coupled conditions. This region of the spectrum is quite unencumbered by pigment absorption, particularly in reaction center preparations, which should make their detection relatively easy. Oddly enough, the only changes that have been measured in the region are characterized by an absorbance *increase* in the region of 450 m$\mu$ [190] rather than a decrease as would be expected for reduction of an iron-sulfur protein.

A further property of iron-sulfur proteins that is inconsistent with their role as the primary electron aceptor is that only those iron-sulfur proteins having more than one iron per molecule have been shown to have an esr signal with a $g$-factor near 1.8 to 1.9 [187-189]. The explanation offered is that in the two-, four-, and eight-iron containing systems the multiple iron atoms are close enough to each other to be antiferromagnetically coupled so that a resultant unpaired electron may be observed only when one unpaired electron still exists between the iron(II) and iron(III) in the cluster. This conclusion is also consistent with recent studies with two well defined model complexes [191, 192] where the structures were determined by x-ray diffraction. The reason why this property seems to be inconsistent with the role of primary electron acceptor is that, by analysis, only one iron atom per primary electron-donor unit was found in reaction center preparations [95]. Consistent with this fact, there were only four cysteine residues and no labile sulfides per primary electron-donor unit (mol wt ~ 75,000). The only well defined iron-sulfur protein that has one iron per molecule and no labile sulfide is rubredoxin [188], and it has no esr signal in the reduced form but a signal at $g = 3.4$ in the oxidized form.

Also bearing on the possible role of iron in the primary photochemical event is the very recent Mossbauer measurement of Debrunner et al. [193] on reaction centers prepared from the R-26 mutant of *R. spheroides*. Using $^{57}$Fe-enriched preparations they studied nonreduced samples and dithionite-reduced samples. It was concluded that iron in both states was in a high-spin ferrous state and that possibly a free radical weakly couples with the iron when dithionite is present.

*4.2.2  Low-Temperature Criteria for an Unchanged Phototrap.* The validity of the use of low temperature measurements concerning the primary photochemical reaction needs to be questioned when applied to physiological temperatures. In an attempt to bridge this gap, McElroy and co-workers [80, 111], Loach et al. [24], and Hsi and Bolton [194] have independently measured the rate of decay of the bacteriochlorophyll cation as a function of temperature from room temperature to liquid helium temperature. Hales and Loach determined this rate for *R. rubrum* chromatophores and photoreceptor subunits (AUT-e) while McElroy et al. and Hsi and Bolton used chromatophores and reaction centers prepared from the R-26 mutant of *R. spheroides*. Each research group found that the first order decay rate increased as the temperature decreased (e.g., see Fig. 22). The "negative activation energy" for this process has been interpreted in terms of a tunneling model for the decay of the electron from the reduced acceptor to the oxidized bacteriochlorophyll donor unit. The increase in rate with decreasing temperature is thus explained as a contraction of the barrier width between the acceptor and donor units causing a decrease in the tunneling half time. It is important to note that each of these studies found close agreement between the decay rates of the chromatophores and purified subunits over the entire temperature range. As discussed below, this is especially significant in the study by Hales and Loach in which the photoreceptor preparation was considered to be almost completely iron free and the apparent electron acceptor is a ubiquinone molecule and not an iron-sulfur protein.

*4.2.3  Low Temperature and Room Remperature Phototrap Activity.* Although the temperature dependency of the dark decay

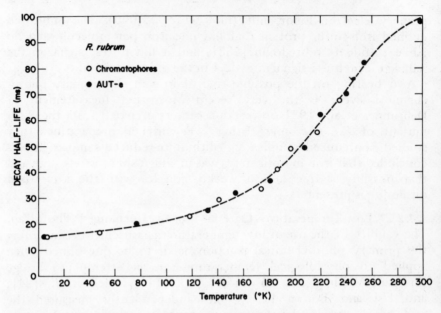

Figure 22. Temperature variation in half-life for the first order decay of the light-induced donor signal in *R. rubrum* chromatophores (○) and AUT-e (●) as recorded by esr spectroscopy. (Taken from [24].)

kinetics is probably a valid means for comparison of *in vivo* and *in vitro* preparations as described above, a very important question to ask is to what extent observations at low temperature apply to the physiological state. There is ample evidence that the phototrap structure changes on cooling, as witnessed by the fivefold increase in decay kinetics (see Fig. 22). A change in structure is also shown by the fact that the bacteriochlorophyll components of the phototrap are significantly modified with respect to the degree of interaction with each other, as is evident upon cooling by a red shift of some 28 nm in the absorbance band whose $\lambda_{max}$ is at 865 nm at room temperature [95]. With regard to the acceptor species, it is possible that iron is forced into a closer proximity to the ubiquinone upon cooling and that the orientation of the ubiquinone becomes fixed in a geometry favoring magnetic interaction with iron. In the case of green plants and algae, lowering the temperature is sufficiently drastic to quench nearly all reversal of phototrap activity. Also in green plants, cytochrome $b_{559}$ becomes a preferred electron donor to the PS II phototrap at low temperature but not at room temperature [243]. Furthermore, from studies of membrane transport systems it is known that phase transitions from a liquid to a gel state associated with lowering the temperature cause concommitant changes in integral protein structure. The functioning of such proteins is thus linked to cooperative changes in macroscopic properties of the membrane lipids [244]. What is badly needed is a method of directly observing the primary electron acceptor under physiological conditions in the intact cell.

*4.2.4* **Esr of an Organic Electron Acceptor.** Working with a purified photoreceptor complex from *R. rubrum*, Loach and Hall [138] first observed and characterized the esr spectrum of a new organic radical that was photoproduced simultaneously with the donor radical. From analytical data it was shown that, of the transition metals present in the bacteria, only iron remained and its concentration was less than 0.3 equivalents per phototrap. The new photoproduced esr signal was detected in these preparations and was shown to be due to a photoreduced species; it had a g-value of 2.0050 ± 0.0003, peak-peak width of 7.5 ± 0.3 G, and a nearly Gaussian shape. This signal also displayed microwave power saturation at lower power levels than did the bacteriochlorophyll donor signal. Furthermore, quantum-yield measurements of spin production showed that the new signal was very efficiently formed ($\Phi$ = 0.6). No hyperfine structure (down to 0.1 G modulation amplitude) was observed in the new signal, either at room temperature or at the temperature of liquid nitrogen. However, structure was observed (presumably from g-anisotropy) in the spectrum of this radical in complexes from bacteria grown in deuterated media [24]. This organic radical was observed later by Feher et al. [195] in reaction centers treated to remove most of the iron. By comparison with model systems, these latter workers identified this organic radical as the anion radical of ubiquinone.

*4.2.5* **Redox Properties at Low Potential.** A major stimulus for reinterpretation of the low-potential titration data of photosynthetic bacteria (e.g., Fig 20) came from results by Loach and Hales [24], who attempted a titration of purified photoreceptor complexes and found that they behaved differently than chromatophores. Subsequent reinvestigation [24] of the low-potential quenching of phototrap activity in whole cells and membrane vesicles showed that the earlier data were not complete. For example, if the biological samples were kept completely dark during the lowering of the environmental potential and then the activity measured by a single, short flash of light, the samples showed activity even at potentials much lower than those previously measured (e.g., at -0.3 V). Redox data from such experiments are shown in Figure 23.*

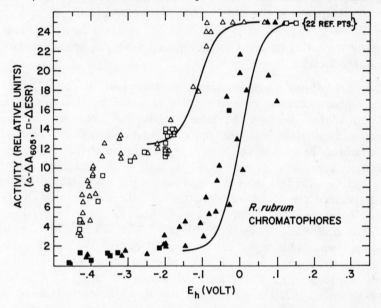

Figure 23. Redox dependency of the first flash light-induced ESR signal (□) or absorbance change at 605 nm (Δ) in chromatophores from *R. rubrum*. Samples were in .05M tris buffer, pH 7.5 in 50% sucrose. Redox buffers used were 2,6 dichlorophenolindophenol, potassium ferrocyanide, indigotetrasulfonic acid, indigotrisulfonic acid, indigodisulfonic acid, flavin mononucleotide (FMN), and Mn (III) Hematoporphyrin IX. All were used at concentrations from $2 \times 10^{-5}$ M to $1 \times 10^{-4}$ M. Typically, $4 \times 10^{-5}$ M was used. The results of many separate experiments are plotted in the figure. All had a reference point at a potential between +.05 and +.30 volt to which the data have been normalized. The solid lines represent one electron titrations with midpoints of 0 and -0.10 volt. Reductant used was $Na_2S_2O_4$. Oxidants used were FMN, potassium ferricyanide and air. The O. D. of the samples for ESR measurements were between 80 and 300, while those for absorbance change at 605 nm were between 2 and 8. Saturating light (visible and near infrared) from a $10\mu$ sec Xenon flash lamp was employed as the exciting light. The redox potential was adjusted in each experiment and the sample taken under dark conditions. Room temperature.

*Note that about half the full activity remains at potentials between -0.1 and -0.35 V rather than the full activity as originally reported. We have recently discovered that an instrumental artifact had not been adequately removed from early measurements [24].

The dark, single-flask redox dependency indicates two kinds of behavior at low potential (below 0 volt). Half the photochemical activity may be reversibly titrated with a midpoint potential of $-0.10$ volt while the other half is only irreversibly lost at much lower potential. In chromatophore preparations from *R. rubrum* it was found that the residual activity (for example, as observed at $-0.25$ volt) would decrease with time as the sample remained at a low redox potential. This loss of activity was greatly accelerated by higher concentrations of FMN (1 hr half-time for loss of phototrap activity at $2 \times 10^{-4}$ $M$ FMN at 25°C and $E_h = -0.20$ V). High viscosity considerably retarded the rate of loss of activity and these titrations were routinely conducted in 50% sucrose. Whole cells were particularly stable at such low potentials in the presence of anionic redox buffers. It should be noted that both whole cells and chromatophores rapidly lost the low potential activity in the presence of cationic viologen dyes.

The above behavior suggests that the primary electron-acceptor molecule may be capable of either a one-electron reduction or a two-electron reduction. (Such a possibility was suggested some years ago [25, 47]). Thus light may normally cause two separate (sequential under normal growing conditions) one-electron reductions of the molecule, whereas a phototrap in thermodynamic equilibration with an added redox buffer may be reduced by two electron equivalents also involving two protons. Further interpretation of the results shown in Fig. 23 are deferred to the section on possible mechanisms of the primary photochemistry.

### 4.2.6 Redox Properties of Ubiquinone.

The oxidation-reduction properties of ubiquinone are consistent with the redox data cited above. In water-containing solvents, one-electron reduction of quinones (at pH 7) is not observed easily because the semiquinone radical quickly disproportionates into two-electron reduced and fully oxidized ubiquinone. The $E_0'$ value for this reaction (two-electron reduction in water) has been reported to be near 0.1 V [196]. However in anhydrous solvents—which are similar to the environment one might expect to find in a membrane or in a hydrophobic region of a protein—one-electron reduced ubiquinone is a stable species [197]; for example, the polarographic half-wave potential for reduction of *p*-benzoquinone in dimethyl formamide is $-0.29$ V versus SHE* and for tetramethyl *p*-benzoquinone it is $-0.50$ V

*Standard hydrogen electrode.

versus SHE. Similarly, the $E_{1/2}$ value of vitamin E quinone is $-0.49$ V and of vitamin $K_1$ is $-0.50$ V versus SHE [197]; the *in vivo* experimental value of $\leqq -0.4$ V is in the potential region expected for the equilibrium involving ubiquinone and its one-electron reduced anion radical in an anhydrous environment.

Also arising from the recent availability of purified reaction centers are a set of light-induced absorbance changes that have been suggested to be associated with the primary electron acceptor. Clayton and Straley [190] carefully documented a small absorbance increase at 450 nm, which they assigned to the absorption spectrum of the reduced primary acceptor. Slooten [198] pointed out the marked similarity of the light-induced absorbance changes observed by Clayton and straley to those for the ubiquinone anion radical minus oxidized ubiquinone as observed by Land et al. [199] in polar solvents (see comparison in Fig. 24). Although these data would appear to be consistent with the esr measurements indicating the formation of ubiquinone anion produced in photoreceptor complexes as outlined previously, Clayton (199a) was unable to find a similar esr signal under conditions where the absorbance change at 450 was produced. Interaction of the ubisemiquinone with an iron-sulfur protein present in the reaction center preparation but not in the photoreceptor pre-

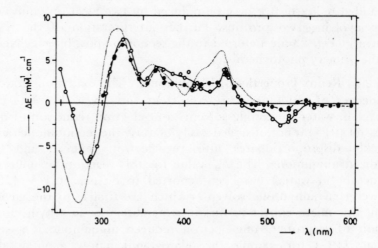

Figure 24.  Light-induced absorbance change ($\bigcirc$ and $\bullet$) observed in bacterial reaction centers compared by Slooten [198] to the absorbance change accompanying the reduction of ubiquinone ($\ldots\ldots$).

parations can be used to explain the discrepency. This is further discussed in a later mechanism section dealing with role of iron.

### 4.2.7 Extractions of Ubiquinone: Tightly Bound Ubiquinone.
Because of the long nonpolar hydrocarbon that is part of the ubiquinone molecule (see Fig. 2), these molecules can be somewhat selectively extracted from membranous fractions without permanently damaging the membrane structure. Many experiments of this sort have been conducted [200-207] and the results have been utilized both to implicate and to rule out a role for ubiquinone. Part of the problem may be that only 0.5 mole of ubiquinone are required per phototrap to maintain full activity (see section 5 in which a mechanism is suggested for the primary photochemical event). This is only one-tenth to one-fiftieth of the total pool size normally present, and most analytical procedures used would not have been sufficiently sensitive to quantitatively measure such a small fraction. particularly if it were covalently bound to protein.

Using a selective labeling technique originally introduced by Rudney and Parson [208-210], we have prepared cells of *R. rubrum*, *R. rubrum* G-9, *R. Speroides*, the R-26 mutant of *R. spheroides, R. capsulata*, and *Chromatium* with highly radioactive [14]C-labeled ubiquinone. Chromatophores were extracted with a number of hydrocarbon solvents and the ubiquinone isolated on silica gel thin layer plates. Losses of ubiquinone during the procedure were accounted for by use of carrier technique. Ubiquinone in the residue that was not extracted with the hydrocarbon solvents was extracted with 1:1 acetone/methanol and determined as indicated above. Finally, the amount of radioactive material resisting all extraction was determined. Analysis of all protein components of chromatophores indicated that less than 0.06 mole of ubiquinone/trap could be covalently bound to protein. The extraction data showed the existence of two separate ubiquinone pools, one easily extractable in hydrocarbon solvents and one remaining tightly bound to chromatophores. The loosely bound pool consisted of 5.0 ubiquinone per trap. Single flash absorbance change data and low temperature light-induced esr signals determined that removal of the loosely bound pool did not diminish the primary photochemistry. When the loosely bound pool was removed from labeled chromatophores and cold ubiquinone added back, little if any exchange occurred between the hot and cold pools. The tightly bound pool had a higher specific activity than the loosely bound pool. Quantitative estimation of the moles of tightly bound ubiquinone per mole primary electron donor unit gave 0.48 with a standard deviation of ±0.06 (see Table 7). This

**TABLE 7   EXTRACTION OF UBIQUINONE[a] AND RHODOQUINONE[b] FROM R. RUBRUM**

| Solvent | Number of Extractions | Time per Extraction (hr) | $UQ^c$ left/trap | $RQ^c$ left/trap |
|---|---|---|---|---|
| Petroleum ether | 1 | 2 | 0.48 | 0.26 |
|  | 2 | 2 | 0.60 | – |
| Petroleum ether +1% acetone | 1 | 17 | 0.48 | – |
| Benzene | 3 | 1 | 0.43 | – |
|  | 1 | 8 | 0.51 | 0.021 |
| Benzene +5% acetone | 1 | 1 | 0.52 | 0.032 |
| Isooctane +1% acetone | 1 | 17 | 0.43 | – |

[a] 6700 CPM ≎ 1 ml of $1 \times 10^{-6}$ $M$ tightly bound ubiquinone.
[b] 8300 CPM ≎ 1 ml of $1 \times 10^{-6}$ $M$ rhodoquinone.
[c] Analysis were performed by extracting the washed residues with 1:1 methanol/ acetone, adding a 100-fold excess of cold carrier, and isolating by thin layer chromatography.

strongly supports the duplex model suggested in the next section and rules out an iron-ubiquinone complex in which ubiquinone is only one electron reduced. It should be emphasized that so far we have only completed such analysis with wild-type R. rubrum.

## 5   POSSIBLE MECHANISMS OF THE PRIMARY PHOTOCHEMISTRY

Before discussion of a specific photochemical reaction it should be appreciated that there may be several excited states of the primary electron-donor unit $[BChl]_4$ that lead sequentially to an appropriately excited electron and result in its capture by another molecule. In addition, one or more transient oxidation states of certain molecules may play a role but remain undetected because of a short lifetime, or our inability to recognize a characteristic state. Perhaps the most important aspect that remains largely unknown is the extent to which the redox properties of, for example, the

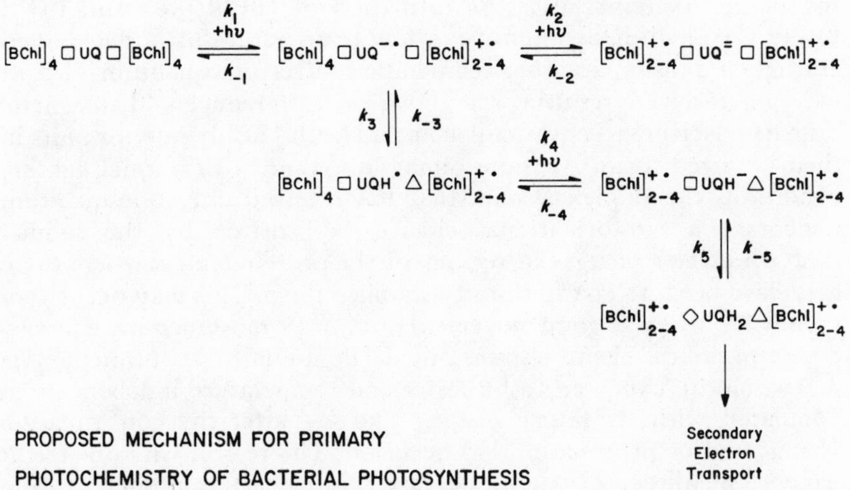

$$[BChl]_4 \,\square\, UQ \,\square\, [BChl]_4 \underset{k_{-1}}{\overset{\overset{k_1}{+h\nu}}{\rightleftharpoons}} [BChl]_4 \,\square\, UQ^{-\bullet} \,\square\, [BChl]_{2-4}^{+\bullet} \underset{k_{-2}}{\overset{\overset{k_2}{+h\nu}}{\rightleftharpoons}} [BChl]_{2-4}^{+\bullet} \,\square\, UQ^{=} \,\square\, [BChl]_{2-4}^{+\bullet}$$

$$k_3 \, \big\Updownarrow \, k_{-3}$$

$$[BChl]_4 \,\square\, UQH^{\bullet} \,\triangle\, [BChl]_{2-4}^{+\bullet} \underset{k_{-4}}{\overset{\overset{k_4}{+h\nu}}{\rightleftharpoons}} [BChl]_{2-4}^{+\bullet} \,\square\, UQH^{-} \,\triangle\, [BChl]_{2-4}^{+\bullet}$$

$$k_5 \, \big\Updownarrow \, k_{-5}$$

$$[BChl]_{2-4}^{+\bullet} \,\lozenge\, UQH_2 \,\triangle\, [BChl]_{2-4}^{+\bullet}$$

$$\downarrow$$

PROPOSED MECHANISM FOR PRIMARY

PHOTOCHEMISTRY OF BACTERIAL PHOTOSYNTHESIS

Secondary
Electron
Transport

Figure 25. Proposed mechanism for the primary photochemical reactions of bacterial photosynthesis. See text for description.

primary electron-acceptor species depend on interaction with the primary electron-donor unit, the protein, iron, or possibly a divalent cation, such as $Mg^{2+}$. Appropriate model chemistry with small molecules has not been conducted. Nevertheless, our knowledge of the *in vivo* system does seem to warrant detailed consideration of the molecular structure at the phototrap site, particularly if such an exercise suggests good experiments that might help prove or disprove a possible mechanism.

## 5.1 Ubiquinone as Primary Electron Acceptor *

The scheme proposed is shown in Figure 25. Unique to the postulate are the multiple oxidation states of quinones that are utilized to allow one quinone molecule to interact with two bacteriochlorophyll donor units. For purposes of following the scheme, a $[BChl]_4 \square UQ \square$ $[BChl]_4$ unit will be referred to as a "duplex system." The box represents the light-harvesting antenna and protein complex that is part of the photoreceptor unit. The box is replaced by a triangle or a diamond to represent a conformational change in the protein of the phototrap. Thus at room temperature the first quantum received by

*A preliminary description of the duplex model has been given (242).

the duplex system causes the formation of $[BChl]_{2-4}^{+\cdot}$ and $UQ^{-\cdot}$. Under very high light intensity at low temperature or under moderate light intensity at room temperature, a second quantum of light may be received resulting in a two-electron-reduced, diamagnetic primary electron-acceptor molecule and both $[BChl]_4$ donor units in their oxidized form. At room temperature and with normal lighting conditions the duplex system that has received only one quantum undergoes a conformational change $(k_3)$ driven by the sudden creation of two charges in regions of the protein matrix where there may have been no counterions to stabilize them. This may occur very rapidly if only small goup movements and/or $H^+$ movements are involved. Return of the electron to the original donor unit (i.e., tunneling) via $k_{-1}$ would be expected to be faster and temperature independent as compared with its return via $k_{-3}$ and $k_{-1}$ after the conformation change and/or protonation has occurred. The reason why one might expect the direct decay to be faster is in part that there are no counterions in the vincinity of the new charges to help stabilize them. The variation in phototrap decay as a function of temperature (see Fig. 22) can thus be explained readily by these equilibria. The decay increases as the temperature is decreased and then becomes essentially invariant between liquid nitrogen and liquid helium temperatures.

As the photoexcitation process continues, the reception of a second quantum of light $(k_4)$ at room temperature under normal lighting conditions would result in loss of an electron by the second donor unit to the semiquinone. This would again be followed by a similar conformational change and/or protonation $(k_5)$ as described above for the $k_3$ step. The end result of the primary photochemistry is thus a diamagnetic, two-electron-reduced quinone and two fully oxidized donor units. The quantum yield for $[BChl]_{2-4}^{+\cdot}$ formation in this scheme would be expected to be near 1.0 as experimentally measured [25, 87].

After the duplex system is completely excited, the reduced electron acceptor may interact with a secondary electron transport component such as an iron-sulfur protein or the ubiquinone or rhodoquinone pool. Secondary electron transport may occur from either the semiquinone form of ubiquinone $(UQH\cdot)$ or the fully reduced form, or from both. Interestingly, an iron-sulfur protein having a two-iron cluster could serve one duplex system as a secondary electron acceptor and still be consistent with the 1:1 iron to $[BChl]_4$ ratio as determined for the carefully documented reaction center complex prepared by Feher [95].

Other experimental observations that are consistent with the scheme are

a. There is no requirement for iron. Photoreceptor preparations from *R. rubrum* are low-iron preparations (some samples had $\leqslant 0.1$ equivalents per phototrap) and yet *in vivo* type of activity could be demonstrated.

b. Under saturating light conditions at room temperature, the only esr signal observed is that due to $[BChl]_{2-4}^{+\cdot}$. One test of the above scheme is to carefully examine the esr signal shape and location under low-intensity light excitation as compared with saturating high-intensity light. If conditions could be found where the radical intermediate was reasonably stable, and low light intensity could be used, the ubiquinone radical species might be detectable unless it interacts with a paramagnetic center such as iron.

c. Procedures that disrupt the membrane structure (e.g., AUT treatment) may physically separate one donor unit from the duplex system, thus allowing the direct observation of $UQ^{-\cdot}$ or $UQH^{\cdot}$ under saturating light conditions. As shown in section 4.2, the semiquinone species was readily demonstrated in photoreceptor preparations. Two other observations with photoreceptor preparations may now become clear: (1) Although the properties of purified photoreceptor complexes (AUT-e) from *R. rubrum* vary somewhat from one preparation to another at most only about half the amount of semiquinone has been observed as $[BChl]_{2-4}^{+\cdot}$ assayed in the chromatophores before treatment. (2) The amount of $[BChl]_{2-4}^{+\cdot}$ obtainable in photoreceptor preparations is always greater in air or with oxidized ubiquinone added in the absence of air than for a simple anaerobic system. This suggests that oxygen or added ubiquinone may allow some $[BChl]_4$ complexes, separated from their original duplex system, to be converted to $[BChl]_{2-4}^{+\cdot}$. With *R. spheroides* AUT-e preparations (active protein-pigment complex) the concentration of $[BChl]_{2-4}^{+\cdot}$ obtained by esr assay anaerobically can often be doubled by adding ubiquinone [139].

d. The amount of $[BChl]_{2-4}^{+\cdot}$ observed as a permanent dark signal at low temperature (for example, at the temperature of liquid nitrogen) depends on how much light was present when the sample was frozen [217]. The higher the light intensity on freezing, the larger the permanent dark $[BChl]_{2-4}^{+\cdot}$ observed. The scheme of Figure 25 nicely explains this observation by the implication that the required conformation changes and/or deprotonation steps ($k_{-5}$ and $k_{-3}$) cannot occur or are frozen in at sufficiently low temperature.

e. A careful comparative study of the ultraviolet- and visible-light-induced absorbance changes in chromatophores of *R. rubrum* and *R. spheroides* and the oxidized minus normal spectra of bacteriochlorophyll in organic solvents showed that the major changes in the *in vivo* system can be ascribed to the photooxidation of the donor unit; however, after subtraction from the *in vivo* absorbance changes of those changes due to oxidation of the donor unit (from BChl oxidation in $CH_2Cl_2$), a small decrease in absorbance at about 280 nm still remains, which is consistent with 0.5 mole (but not 1.0) mole of ubiquinone becoming reduced per BChl]$_4$ (eq. 5).

$$UQ + 2e + 2H^+ \rightarrow UQH_2 \quad (\text{or} \quad UQ^{2-} \quad \text{or} \quad UQH^-) \qquad (5)$$

f. The dependency of phototrap activity on redox potential at low potential and under dark, single pulse conditions (Fig. 23) have shown that only half the activity is quenched with a midpoint of −0.10 volt and the other half is maintained to very low potential. Whereas in the presence of background illumination, or upon subsequent oxidative titration after very low potential was reached in the dark, single flash experiments (Fig. 23), the midpoint potential is −0.02 volt. With a consideration of ubiquinone oxidation-reduction chemistry, these data may be understood as follows. The reversible redox titration for the first half of the phototrap quenching (indicated in Fig. 23 by the change having an apparent midpoint of −0.10 volt) may be interpreted as due to the equilibria

$$H^+ + [BChl]_4 \square UQ \square [BChl]_4 + e \rightleftharpoons [BChl]_4 \square UQH \cdot \square [BChl]_4$$

As an alternative to stabilization of the free radical by hydrogen ion, a positive charge from the protein in the region of $UQ^{-\cdot}$ might also stabilize this species.

Quenching of the second half of the phototrap activity at very low potential would then be ascribed to the addition of a second electron with or without protons. This reaction shows a hysteresis effect which is characteristic of membrane cooperative systems and suggests that a major change occurs either in the macroscopic properties of the membrane (e.g., a change in fluidity, Trauble and Eibl, 1975), or of the protein. By this model, a reason why the free radical species of the primary electron acceptor is not observed in *in vivo* systems (except perhaps at concentrations lower than about 10% of the phototrap complex) could be because the usual lighting conditions drive the phototrap complex to its fully charged condition in which ubiquinone is two electron reduced and therefore, diamagnetic.

The hysteresis effect may be explained by further consideration of ubiquinone oxidation-reduction chemistry. The quenching of all activity at lower potential may be the result of the reaction

$$H^+ + [BChl]_4 \square UQH \cdot \square [BChl]_4 + e \rightleftharpoons [BChl]_4 \square UQH_2 \square [BChl]_4$$

which may occur by direct reduction

$$[BChl]_4 \square UQH \cdot \square [BChl]_4 + e \rightleftharpoons [BChl]_4 \square UQH^- \square [BChl]_4$$

followed by protonation. The latter reaction would require a very strong reductant for it to proceed spontaneously to the right. However, addition of a proton greatly stabilizes the $UQH^-$. Once formed, the $UQH_2$ would not be reoxidized until sufficiently high potential where both an electron and a proton (or two electrons and two protons) could be removed. One electron oxidation without loss of a proton is highly unfavored ($E_{1/2} \cong 1.1$ volt; see Table 8).

$$[BChl]_4 \square UQH_2 \square [BChl]_4 \rightleftharpoons [BChl]_4 \square UQH_2^+ \cdot \square [BChl]_4 + e$$

Reoxidation of the $UQH_2$ in the phototrap may depend on reoxidation of the first equivalent of the UQ pool whose molecules may in turn interact with the phototrap $UQH_2$, in a redox reaction involving two electrons and two protons. This would be the transition with a midpoint of $-0.02$ volt. Thus the expected oxidation-reduction chemistry of ubiquinone nicely explains the observations.

## 5.2 Appreciation of Multiple Photochemical Reactions Possible between Bacteriochlorophyll and Ubiquinone

As a result of recent progress in porphyrin chemistry, the oxidation states and protonation states shown in Table 9 could reasonably be

### TABLE 9   POSSIBLE REDOX STATES OF THE PRIMARY ELECTRON-DONOR UNIT

| | | |
|---|---|---|
| $[BChl]_n^{2+}$ | $[BChl]_n(OH)^+$ | $[BChl]_n(OH)_2$ |
| $[BChl]_n^{+\cdot}$ | $[BChl]_n(OH)^\cdot$ | |
| $[BChl]_n$ | | |
| $[BChl]_n^{-\cdot}$ | $[BChl]_n H^\cdot$ | |
| $[BChl]_n^{2-}$ | $[BChl]_n H^-$ | $[BChl]_n H_2$ |

## TABLE 8 OXIDATION-REDUCTION BEHAVIOR OF BACTERIOCHLOROPHYLL, BACTERIOPHEOPHYTIN, AND QUINONES[a]

| Redox Couple | Solvent | $E'_0{}^b$ (V) | $E_{1/2}{}^b$ (V) | References |
|---|---|---|---|---|
| BChl$^{+\bullet}$/BChl | Methanol(anhydrous) | 0.27 | | 56, 213 |
| | Methanol, 1% $H_2O$, pH 7.0 | 0.00 | | 213 |
| BPh$^{+\bullet}$/BPh | Dichloromethane | | 0.40 | 214 |
| BChl$^{2+}$/BChl$^{+\bullet}$ | Dichloromethane | | 0.72 | 214 |
| BChl/BChl$^{-\bullet}$ | Methanol | 0.59[c] | | 215 |
| BPh/BPh$^{-\bullet}$ | Butyronitrile | | $-1.08$ | 215 |
| | — | | $-0.8$[d] | |
| Mg(II)TPP/Mg(II)TPP$^{-\bullet}$ | Dimethyl sulfoxide | | $-1.35$ | 216 |
| H$_2$TPP/H$_2$TPP$^{-\bullet}$ | Dimethyl sulfoxide | | $-1.05$ | 216 |
| H$_2$ Etioporphyrin I/H$_2$ Etioporphyrin I$^{-\bullet}$ | Dimethyl formamide | | $-1.34$ | 216 |
| Mg(II)TPP$^{-\bullet}$/Mg(II)TPP$^{2-}$ | Dimethyl sulfoxide | | $-1.80$ | 216 |
| H$_2$TPP$^{-\bullet}$/H$_2$TPP$^{2-}$ | Dimethyl sulfoxide | | $-1.47$ | 216 |
| BChl/BChlH$_2$ | Methanol, 1% $H_2O$, pH 7.2 | $-0.42$ | | 217 |
| BPh/BPhH$_2$ | Methanol, 1% $H_2O$, pH 7.2 | $-0.11$ | | 217 |
| | Methanol (anhydrous) | $-0.06$ | | 217 |
| BQ/BQ$^{-\bullet}$ | Dimethyl sulfoxide | $-0.40$ | | 197 |
| | Dimethyl formamide | $-0.54$ | | 197 |
| | Acetonitrile | $-0.52$ | | 197 |

| | | | |
|---|---|---|---|
| 2,5-dimethyl BQ/2,5-diMethyl BQ⁻· | Acetonitrile | −0.54 | 197 |
| 2,5-dimethoxy BQ/2,5-dimethoxy BQ⁻· | Dimethylformamide | −0.7 | 197 |
| BQ⁻·/BQ²⁻ | Dimethyl sulfoxide | −1.2 | 197 |
| 2,5-dimethoxy BQ/2,5-dimethoxy BQ²⁻ | Dimethylformamide | −1.3 | 197 |
| UQ/UQH₂ | | 0.1 | |
| BQH₂⁺·/BQH₂ | Acetonitrile | 1.1 | 197 |

[a] For abbreviations, see Section 7.
[b] All values reported are relative to the saturated calomel electrode (SCE).
[c] Estimated to be 0.32 V above the $E'_0$ value of the monocation [59].
[d] Estimated by assuming that the difference between BChl and BPh would be the same as that between MgTPP and TPP.

considered to play a role in photosynthesis. There are 11 states and 16 possible one-electron equilibria. Similarly, Table 10 shows relevant

### TABLE 10   POSSIBLE REDOX STATES OF UBIQUINONE

| | | |
|---|---|---|
| $UQ$ | $UQH^+$ | $UQH_2^{2+}$ |
| $UQ^{-\cdot}$ | $UQH^{\cdot}$ | $UQH_2^{+\cdot}$ |
| $UQ^{2-}$ | $UQH^-$ | $UQH_2$ |

redox and protonation states that are reasonable to consider for ubiquinone. In this case there are 9 states and again 16 possible one-electron equilibria. The number of one-electron photochemical reactions possible by combining these various pairs approaches 100,000! Equation 6 is just one example of a reaction that could be considered theoretically. Fortunately, this particular reaction can be

$$[BChl]_4 \square UQH_2 \xrightarrow{+h\nu} [BChl]_4^{-\cdot} \square UQH_2^{+\cdot} \qquad (6)$$

ruled out (at least for bacterial photosynthesis) because insufficient energy is available in a quanta of 880 nm light to drive the reaction (see Table 8 for appropriate $E_{1/2}$ values). However the example serves to underscore the possible events to consider when both the primary electron-donor and the primary electron-acceptor have multiple redox states available and excited prophyrin molecules are known to be either exceedingly good electron donors or acceptors. In the event that $[BChl]_{2-4}^{+\cdot} \square UQ^{-\cdot} \square [BChl]_4$ is not equivalent to $[BChl]_4 \square UQ^{-\cdot} \square [BChl]_{2-4}^{+\cdot}$, a further complication of additional possible states would exist.

Obviously, the biological systems have evolved over the years to have presumably selected that particular reaction that best suits their purpose. It is informative at this time to ask how a person attempting to understand photosynthesis by a purely model approach knows which of the possibilities outlined above to select and, furthermore how to limit the reactions that can occur?

Fortunately, for the *in vivo* system we can probably eliminate most of the possible photoreactions projected above by observing a few facts and making a few reasonable assumptions. The ones I would like to consider here are as follows:

1.   Reactions involving doubly oxidized or double reduced bacteriochlorophyll can be ignored because the light energy would have to be absorbed by the one-electron-oxidized or the one-electron-

reduced species and they do not have an appropriate absorbance band in the near infrared to accept excitation energy from the antenna complex.

2. Only redox equilibria that do not involve a change in the protonation state are likely to occur in the photochemical reaction.

3. To the extent that they are applicable, electrochemical data can provide a limit for reactions possible on energetic grounds.

After applying these three restrictions, only the five possibilities given by eqs. 7-11 seem to remain.

$$[BChl]_4 \square UQ \xrightarrow{+h\nu} [BChl]_{2-4}^{+\bullet} \square UQ^{\bullet} \qquad (7)$$

$$[BChl]_4 \square UQ^{\bullet} \xrightarrow{+h\nu} [BChl]_{2-4}^{+\bullet} \square UQ^{2-} \qquad (8)$$

$$[BChl]_4 \square UQH^{\bullet} \xrightarrow{+h\nu} [BChl]_{2-4}^{+\bullet} \square UQH^{-} \qquad (9)$$

$$[BChl]_4 \square UQ^{-\bullet} \xrightarrow{+h\nu} [BChl]_{2-4}^{-\bullet} \square UQ \qquad (10)$$

$$[BChl]_4 \square UQ^{2-} \xrightarrow{+h\nu} [BChl]_{2-4}^{-\bullet} \square UQ^{-\bullet} \qquad (11)$$

Reactions 7-9 are indeed those incorporated into the scheme proposed. Reactions 10 and 11 are very interesting and, according to solution photochemistry, would be very likely to occur. The fact that they do not occur to any significant extent (at least in bacterial photosynthesis) is required because the quantum yield for the secondary reaction (eq. 12) is near 1.0 [25, 47]. In fact, the

$$Fe^{2+} cyt\ c_2 + [BChl]_{2-4}^{+\bullet} \rightleftharpoons Fe^{3+} cyt\ c_2 + [BChl]_4 \qquad (12)$$

subsequent protonation of UQ⁻ may be an essential step to prevent this radical from acting as an electron donor in subsequent light events. The reaction given by eq. 13 would require almost twice

$$[BChl]_4 \square UQH^{\bullet} \xrightarrow{+h\nu} [BChl]_{2-4}^{-\bullet} \square UQH^{+} \qquad (13)$$

the amount of energy available in bacterial photosynthesis. An interesting consequence of the foregoing consideration is the fact that at low temperature if one assumes changes in protonation cannot occur, the reaction measured under saturating light conditions may give eq.

$$[BChl]_{2-4}^{+\bullet} \square UQ \square [BChl]_{2-4}^{-\bullet} \qquad (14)$$

14 rather than eq. 15. Esr spectroscopy, in its current state of

$$[BChl]_{2-4}^{+\bullet} \square UQ^{2-} \square [BChl]_{2-4}^{+\bullet} \tag{15}$$

development, probably could not distinquish between these alternatives.

## 5.3   Duplex VS. Simplex Model

The reasoning that leads to the duplex system for the phototrap rather than favoring one in which only one donor unit and one ubiquinone are required deserves further comment. The latter model for the phototrap may be referred to as a "simplex" model. In the simplex model, I would suggest that the same multiple redox states of ubiquinone may play an important role, with the major difference being that energy must be funneled sequentially into only one donor unit.

The question might be raised as to why an organism—presumably driven by the pressures of survival—would evolve a *seemingly* more complex duplex system rather than a simplex system. I submit that the duplex may in fact be simpler than the simplex when the entire photosynthetic unit is considered. For example, a duplex system effectively expands by a facter of 2 the number of antenna pigments that directly feed a single phototrap and therefore a single acceptor site. A distinct saving on the expenditure of metabolic energy for biosynthesis of secondary electron-transport components (and possibly   coupling site components for photophosphorylation) would seem to represent a major guiding principle for evolutionary adaptation. The duplex system has this advantage, plus all those of the simplex system. It is of course conceivable that some bacteria or mutants do utilize a simplex system.

Another possible explanation of the result of finding 0.5 mole ubiquinone per primary electron donor unit, $BChl_4$, would be that there are two types of phototraps or reaction centers, only one of which utilizes ubiquinone as the primary electron acceptor. The existance of multiple photosystems in bacteria is an old idea, but little evidence exists to support this possibility, particularly in *R. rubrum*. In fact, it would be very difficult to explain experimental results such as a single pseudo-first order decay of the light-excited phototrap at low temperature if two different phototrap populations having different primary electron acceptors were actually present.

## 5.4  Role of the Iron-Sulfur Protein

It has been suggested [195, 235, 236] that iron may be very near the electron acceptor so that it interacts with the latter in its free radical form. A few as yet unexplained observations are: (a) demonstration of an esr signal at low temperature with $g = 1.8$ (e.g., like that of a two-iron cluster such as in spinach ferredoxin); (b) upon illumination, or chemical reduction, purified photoreceptor complexes exhibit a stable ubiquinone anion radical species by esr that is not futher reduced either by light or chemical reductants; (c) in dark, single-flash experiments at low potential conducted with whole cells and chromatophores, only half the amount of $[BChl]_{2-4}^{+\cdot}$ observable above 0 V can be generated in the single flash (unpublished results and Fig. 23). (d) Although changes in absorbance consistent with the formation of ubisemiquinone have been characterized by Clayton and Straley [190], the expected esr signal from the organic radical was not found.

One explanation has already been offered above, in part. That is, for the reaction center preparations studied by Feher, the original duplex may have been broken during isolation of the reaction center and only one electron reduction of ubiquinone can occur because only one donor unit remains with the complex. This unpaired electron then either normally interacts with the iron center, or is forced into a location near the iron center upon freezing. Recall that the tunelling half time shortens five fold as the temperature is lowered, perhaps bearing witness to significant structural rearrangement.

For a second possible explanation, consider the equilibrium given by

$$[BCHl]_4 \square UQ \square [BChl]_4 \quad +e \; + \; H^+ \; \rightleftharpoons \; [BChl]_4 \square UQH_i^{\cdot} \; \square [BChl]_4 \quad (16)$$

$$(Fe)_2^{\cdot} \qquad\qquad\qquad (Fe)_2^{\cdot}$$

where $(Fe)_2^{\cdot}$ represents an interacting iron-sulfur prosthetic grroup [two iron center, one Fe(II), and one Fe(III) which are antiferromagnetically coupled]. Suppose that the $E_0'$ for this reaction is about $-0.05$ V, which is somewhat higher than expected because of the stabilizing effect of the iron on the ubiquinone radical. Thus absorption of a quantum of light in the semiquinone redox state might give $[BChl]_4 \square UQH^- \square BChl_{2-4}^{+\cdot}$. This would not require an

$$(Fe)_2^{\cdot}$$

oxidation state change in the iron-sulfur center [185, 193] but would allow the latter to become esr observable upon illumination because of full reduction of the ubiquinone, thus explaining point $a$ above. Point $c$ would follow because the first electron was chemically added to ubiquinone in the dark as the redox potential was lowered below $-0.05$ V. Therefore only half of the donor units could give up an electron to the ubisemiquinone upon illumination. Point $b$ is satisfied because preparation of photoreceptor complexes both breaks the duplex and removes the iron. When reduced, the anion radical of ubiquinone is the form observed [138, 195], which is consistent with the fact that the $pK_a'$ for protonation of quinone radical anions is about 5 [229, 230]. As a result. the anion is stable in the aqueous phase, whereas the uncharged form, or potonated form, is much more stable in the nonaqueous phase of the membrane. No significant full reduction of the ubiquinone anion radical would occur because formation of the dianionic species requires a much stronger reductant than is available, and disproportionation of ubisemiquinone only takes place between the protonated molecules, which are present only in low concentration and also are very slow to diffuse because of their tight binding to the high molecular weight photoreceptor complex. And finally, point $d$ could be explained by the fact that the reaction center preparations in question still retained the iron-sulfur protein. Thus conditions were found under which a steady state concentration of ubisemiquinone was observed, but the organic radical could not be seen by esr because of coupling to the iron-sulfur center.

The presumed physiological role for the iron-sulfur protein would be that of a secondary electron-transport component with an $E_0'$ value in the range of $-0.05$ to $+0.25$ V for the reaction

$$Fe(III), Fe(III) + e \rightleftharpoons Fe(II), Fe(III) \qquad (17)$$

## 5.5   A Question of Semantics: Pseudoprimary Electron Acceptors

Our understanding of the primary events in photosynthesis is approaching a time when a more precise way of discussing the chrological events is required. The primary electron-donor and the primary electron-acceptor species have been most often defined as that species that first gives up or accepts an electron, respectively, after a quantum of light energy is absorbed. To apply this definition in a practical way, it is often assumed implicitly that the species formed have a long enough life time so that they can be detected under physiological conditions. As the capability for measuring changes in oxidation state has progressed to shorter and shorter times, we now find ourselves on a time scale as short as the lifetime of a variety to excited states. At what point, for example, does one cease to talk about a triplet state and begin to talk about on ion pair, or a reduced and an oxidized species. This is made more difficult by the recognition that the phototrap contains six porphyrin analogs in close proximity to each other, at least two of which function as a unit.

Perhaps an additional criterion should be added; for example, the redox properties of the primary electron-acceptor species should be unaffected by the oxidation state of the primary electron donor. This would provide a distinction between excited states and redox states of independent entities. Following the path of an electron from an excited state to ubiquinone might involve several intermediate reduced species, which, for the purpose of discussion, I will call pseudoprimary electron acceptors. The possible events of interest are given by eq. 18.

$$\left\{ [BChl_2BChl_2]\,Bph_2 \right\}\ \Box UQ \xrightarrow{+h\nu} \left\{ [BChl_2^{+\bullet}BChl_2^{-\bullet}]\,Bph_2\ \Box UQ \right\}$$

$$\downarrow$$

$$\left\{ [BChl_2^{+\bullet}BChl_2]\,Bph_2^{-\bullet}\ \Box UQ \right\} \qquad (18)$$

$$\downarrow$$

$$\left\{ [BChl_2^{+\bullet}BChl_2]\,Bph_2\ \Box UQ^{-\bullet} \right\}$$

The energetics of this process are particularly interesting. In the first step creation of a bacteriophlorophyll cation radical and a bacteriochlorophyll anion radical requires an energy essentially equal to that available in a quantum of light at 880 nm (see Table 8). Since theory predicts [218-220] that about 35% of the energy available in the original quantum must be lost to the surroundings for the products to be stable, a very short lifetime is predicted for such an ion pair. In the next step approximately 20% of the energy originally available may be lost as the bacteriopheophytin anion radical is formed (again, see Table 8). Finally, as the ubiquinone anion radical is formed, another 15% of the energy is lost and the products have conserved about 65% of the original energy in the quantum absorbed. This process could be referred to as an energy "cascade," which also may result in a significant separation of the charged species in space. Could such an involvement explain the presence of bacteriopheophytin in the phototrap? Recent experimental results measuring very short-lived species are consistent with this interpretation (241).

## 5.6  Two Photosystems

The mechanism of the primary photochemical event proposed in Figure 25 has several attractive features for extension to photosynthetic organisms having two different kinds of photosystems. One part of the duplex system may have changed with time with regard to the supporting light-harvesting antenna pigments and the interaction with secondary electron-transport components; for example, one could imagine an extension to green plants and algae where two chlorophyll aggregates (perhaps a special pair [106]) serve as the primary electron-donor unit called P700. This scheme is shown in Figure 26. The donor unit on the right half of the duplex would be the system I P700 and that on the left would be the system II P700. It would not change the model substantially to make one donor unit (the one on the left of the duplex) P680 [221, 222], but the

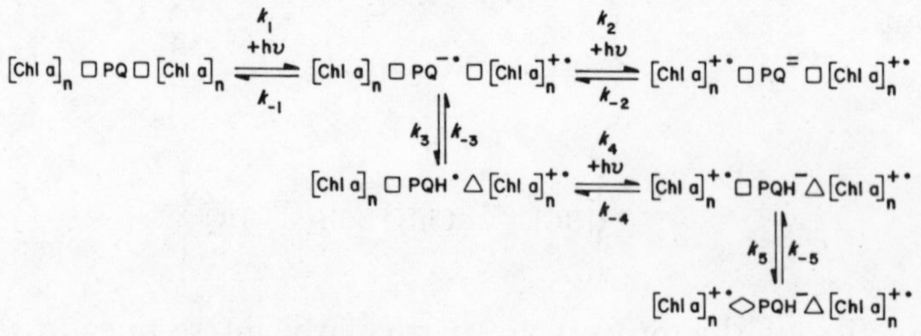

$$[Chl\ a]_n\ \Box\ PQ\ \Box\ [Chl\ a]_n \underset{k_{-1}}{\overset{\underset{+h\upsilon}{k_1}}{\rightleftharpoons}} [Chl\ a]_n\ \Box\ PQ^{-\cdot}\ \Box\ [Chl\ a]_n^{+\cdot} \underset{k_{-2}}{\overset{\underset{+h\upsilon}{k_2}}{\rightleftharpoons}} [Chl\ a]_n^{+\cdot}\ \Box\ PQ^=\ \Box\ [Chl\ a]_n^{+\cdot}$$

$$\Big\updownarrow k_3 \Big\| k_{-3} \qquad\qquad\qquad k_4$$

$$[Chl\ a]_n\ \Box\ PQH^\cdot\ \triangle\ [Chl\ a]_n^{+\cdot} \underset{k_{-4}}{\overset{\underset{+h\upsilon}{k_4}}{\rightleftharpoons}} [Chl\ a]_n^{+\cdot}\ \Box\ PQH^-\ \triangle\ [Chl\ a]_n^{+\cdot}$$

$$\Big\updownarrow k_5 \Big\| k_{-5}$$

$$[Chl\ a]_n^{+\cdot}\ \diamond\ PQH^-\ \triangle\ [Chl\ a]_n^{+\cdot}$$

1.  $[Chl\ a]_n^{+\cdot} \diamond PQH^\cdot \triangle [Chl\ a]_n^{+\cdot} + Fe_{NHI}^{3+} \rightleftharpoons [Chl\ a]_n^{+\cdot} \diamond PQH^- \triangle [Chl\ a]_n^{+\cdot} + Fe_{NHI}^{2+}$

2.  $[Chl\ a]_n^{+\cdot} \diamond PQH^\cdot \triangle [Chl\ a]_n^{+\cdot} + Mn(H_2O)Cl_2 \rightleftharpoons [Chl\ a]_n \Box PQH^\cdot \triangle [Chl\ a]_n^{+\cdot} + Mn(H_2O)(OH)Cl_2 + H^+$

3.  $[Chl\ a]_n \Box PQH^\cdot \triangle [Chl\ a]_n^{+\cdot} + Fe_{NHII}^{3+}$ or $PQ \rightleftharpoons [Chl\ a]_n \Box PQ \Box [Chl\ a]_n^{+\cdot} + Fe_{NHII}^{2+}$ or $PQH^\cdot$

4.  $[Chl\ a]_n \Box PQ \Box [Chl\ a]_n^{+\cdot} + Fe^{2+}Cyt\ f$ or $Cu^+PC \rightleftharpoons [Chl\ a]_n \Box PQ \Box [Chl\ a]_n + Fe^{3+}Cyt\ f$ or $Cu^{++}PC$

5.  $3[Chl\ a]_n^{+\cdot} \diamond PQH^\cdot \triangle [Chl\ a]_n^{+\cdot} + Mn(H_2O)(OH)Cl_2 \rightleftharpoons 3[Chl\ a]_n \Box PQH^\cdot \triangle [Chl\ a]_n^{+\cdot} + Mn(O^=)_2Cl_2 + 3H^+$

6.  $Mn(O^=)_2Cl_2 + 2H_2O \rightleftharpoons Mn(H_2O)_2Cl_2 + O_2$

PROPOSED MECHANISM FOR OXYGEN EVOLVING ORGANISMS

Figure 26.  Proposed extension of the Duplex model to oxygen-evolving systems. See text for description and the list of abbreviations at the end of the chapter.

evidence is still weak with regard to a role of P680 as the primary electron donor.

As can be seen in eq. 1 of Figure 26, fully reduced plastoquinone may first react with a ferredoxin type of secondary electron transport component. $PQH^-$ would be a strong reductant, probably a little stronger than $PQ^-$, and would be consistent with the reducing capacity of the "system I primary electron acceptor." This reducing equivalent would then make its way through other electron carriers to $NADP^+$. The remaining $PQH^.$, which is still a good reductant, would then interact with a different secondary electron-transport component (Fig. 26, eq. 3), with the reducing equivalent eventually going to the PQ pool and the interconnecting part of the traditional "Z scheme."

In the meantime, when both P700 complexes are oxidized, the oxidizing potential for one equivalent could well be significantly higher if a membrane potential also exists or if the second $P700^{+.}$ is in close proximity to the first. As indicated in Figure 26 (eq 2), the first oxidizing equivalent is viewed as being sufficiently strong to oxidize the redox center at the oxygen-evolving site. Subsequent excitation of the system II P700 results in sequential storage of oxidizing equivalents (Fig. 26, eq. 5) and eventual oxygen evolution (Fig. 26, eq. 6). A manganese chloride complex is invoked on purely speculative grounds. Once the first oxidizing equivalent is passed on to the secondary electron-donor component, the remaining $P700^{+.}$ is still a strong oxidant but with properties like those usually assigned to the system I primary electron donor (Fig. 26, eq. 4). There is some experimental evidence consistent with two populations of P700 having different redox properties [55, 223, 224]. Many interesting experimental results fit this scheme, and some appear inconsistent.

In the latter category, Ke has provided evidence that the primary electron acceptor of system I may have an absorbance change at 430 nm and has called the component P430 [237-239]. The absorbance change upon illumination is a decrease and is consistent with the reduction of an iron-sulfur protein. Since several iron-sulfur protein components can be observed at low temperature and appear to be reduced upon illumination [239], Ke has suggested that P430 is a membrane-bound iron-sulfur protein and serves as the primary electron acceptor in photosystem I. I would like to suggest an alternative interpretation of the absorbance changes at 430 nm that could, in fact, be used to support the duplex model*. Experimentally

*A preliminary description of the application of the duplex model to green plant photosynthesis has been given (247).

the change at 430 nm has been shown to consist of two parts: (a) a decrease in absorbance upon illumination due to the oxidation of P700 that recovers in the dark and (b) a kinetically distinct portion of the absorbance increase during dark recovery at 430 nm used to distinguish P430 from P700. Also, the kinetic response to added redox components of the portion ascribed to P430 was shown to be consistent with a reduced species. With regard to the duplex model, PQ would presumably change from completely oxidized to completely reduced (2 electrons + 1 proton) during illumination, a process that would have no effect on the absorbance change at 430 nm. However, reduced plastoquinone would then be reoxidized in the subsequent dark period and, depending on the rate of delivery of each equivalent of electrons to secondary electron acceptors, would first uniquely form the free radical of plastoquinone and then subsequently become completely oxidized. The free radical would be expected to have an absorbance band in the vicinity of 430 nm (c.f., Fig. 24). Alternatively, the free radical form of PQ may have been present before light excitation and may have been interacting with an iron-sulfur center as suggested earlier for the photosynthetic bacteria; upon excitation it would become further reduced, thus resulting in absorbance changes like those observed and allowing the iron-sulfur center to become esr detectable. Thus what has been interpreted as a restoration of an absorbance band of an iron-sulfur protein as it becomes reoxidized could in fact be the formation of a plastoquinone radical species from fully reduced plastoquinone.

Therefore the same basic photosynthetic unit as is still functional in the photosynthetic bacteria may be used in oxygen-evolving organisms with the following modifications: (a) Expansion and differentiation of the light-harvesting antenna pigments. (b) Evolution of an enzyme complex (possibly involving manganese and chloride) to enable storage of oxidant and catalysis of water oxidation. (c) Selection of enzymes, or electron carriers, that are able to get the electron(s) out of the primary electron acceptor ($PQH^-$) at low potential.

A somewhat different extension to two photosystems of the duplex model would be to consider the phototrap of each separate photosystem as a duplex. System I would then function in a manner very similar to the bacterial system. Using a duplex system for the photosystem II phototrap would still have the advantage of providing two equivalents of oxidant, one at higher potential, and two equivalents of reductant [225].

# 6 CONCLUDING REMARKS

In this chapter I have attempted to present a summary of experimental results that have largely been responsible for our present understanding of the primary photochemical event in bacterial photosynthesis. I have focused attention on the primary electron-donor unit, the protein, the membrane, the primary electron-acceptor species, and the mechanism of the primary photochemical event in that order. The extent to which experimental data supports the discussion decreases approximately in the same order.

All bacterial phototraps seem to consist of four bacteriochlorophyll molecules, two bacteriopheophytin molecules, and a molecule, or half a molecule, of ubiquinone, all of which are specifically bound by a protein of mol wt = 45,000 to 70,000. At least two, and perhaps all four, of the bacteriochlorophyll molecules act together as a supramolecular complex both in absorbing light energy and in sharing the oxidizing equivalent after an electron is donated to the acceptor species. This molecular aggregate appears to have evolved to allow significant overlap of the intensely absorbing near infrared band (often near 870 nm) of this unit with the long-wavelength band of the even more extensively aggregated antenna complex (which absorbs at 880 nm in *R. rubrum*). Model studies with purified chlorophyll and bacteriochlorophyll have shown that self aggregation (which allows some $\pi$-overlap of the extensively unconjugated macrocyclic systems) can promote such extensive red shifts [63-73, 152]. Interaction of monomeric bacteriochlorophyll with protein or phospholipid does not result in such a marked band shift [226]. Utilization of a supramolecular complex for the primary electron-donor unit may also be advantageous for the oxidized equivalent to be extensively delocalized, since this might allow the secondary redox reaction of this species with ferrous cytochrome $c_2$ to occur at some distance from the primary electron-acceptor site.

Successful isolation of relatively small photoreceptor and reaction center complexes has heralded the onset of the application of biochemistry to the study of primary events in photosynthesis. The best defined preparations seem to be the photoreceptor complex isolated from *R. rubrum,* which has two major polypeptides associated with it (apparent mol wt = 19,000 and 32,000), and the reaction center complex isolated from the R-26 mutant of *R. spheroides*, which also has two polypeptides associated with it (apparent mol wt = 24,000 and 21,000). The former preparation retains antenna pigments while the latter does not. The individual polypeptides from each of these preparations have been isolated in sufficient quantities to allow determination of amino acid composition, and amino acid sequence work has begun on one of them. Reconstitution of active complexes from the well characterized purified components is the goal of the next 5 to 10 years. It should be emphasized that the protein characterization work is still at a primitive stage because of the difficulties associated with studying a membrane system. For example, all laboratories do not yet agree even on the minimal number of polypeptides required for a fully functional phototrap or photoreceptor complex.

A summarizing view (as interpreted by our laboratory) of the photoreceptor complex and the membrane was presented in Figure 19. It is important to note that from 50 to 75% of the membrane protein in photosynthetic bacteria is accounted for by the protein of the photoreceptor complexes. Of this amount, in *R. rubrum* over half is acounted for by a single polypeptide of mol wt = 19,000. This particular polypeptide is soluble in organic solvents, which will make structural and comparative studies particularly interesting.

Although it is not generally accepted that ubiquinone is that molecular species that serves as the primary electron acceptor in bacterial photosynthesis, I feel that the evidence summarized in this chapter is compelling and I have devoted little discussion to alternative possibilities. A consideration of the variety of one-electron photochemical reactions that are possible between bacteriochlorophyll and ubiquinone is instructive in pointing out the unlikeliness of discovering the mechanism of photosynthesis by merely selecting a model system without first appreciating the nature of the high selectivity imposed by the protein of the living system. It appears that it would be most profitable to better define the actual course of the photochemical reaction in *in vivo* photosynthesis and then to construct models to confirm the mechanism.

All facts known to me are compatible with a mechanism whereby a single ubiquinone molecule serves as an electron acceptor to two bacteriochlorophyll donor units in two separate light events (see Fig. 25). In fact, this is the only scheme that seems to be consistent with *all* experimental results. It has been presented here as a hypothesis that is readily testable.

# 7 ABBREVIATIONS AND SYMBOLS

## 7.1 Pigments and Redox Components

| | |
|---|---|
| BChl | Bacteriochlorophyll (see Fig. 2 for structure). |
| $[BChl]_4$ | The four molecules of the bacterial phototrap that interact specifically and compose the primary electron-donor unit |
| P870 | A term often used [90, 91] in the literature for $[BChl]_4$. This species often has an absorbance band at 870 nm |
| $P._{44}$ | Another term that has been used [38, 47, 55] to represent $[BChl]_4$. This species usually has an oxidation-reduction midpoint potential of +0.44 V |
| $P_1$ | General symbol used in this chapter for the primary electron-donor species. In *R. rubrum* it would be $[BChl]_4$ |
| $[BChl]_{2\text{-}4}^{+\cdot}$ | The one-electron-oxidized primary electron-donor unit in bacterial photosynthesis. At least two, and perhaps all four, bacteriochlorophyll molecules share the cation radical |
| BPh | Bacteriopheophytin; bacteriochlorophyll without magnesium in the center |
| $[BPh]_2$ | The two bacteriopheophytin molecules so far found in all bacterial phototraps. No special interaction or aggregate is intended |
| $[Chl]_n$ | The $n$ molecules of the phototrap that interact specifically and compose the primary electron-donor unit of system I (perhaps also of system II) |
| P700 | A term often used for $[Chl]_n$ especially as applied to system I. This species often has an absorbance band at 700 nm |
| $[Chl]_{2\text{-}n}^{+\cdot}$ | The one-electron-oxidized primary electron-donor unit in oxygen-evolving organisms |

| | |
|---|---|
| Ph | Pheophytin $a$; chlorophyll $a$ without magnesium in the center |
| UQ | Ubiquinone (for structure, see Fig. 2) |
| $[UQ]_n$ | A pool of $n$ molecules of ubiquinone |
| PQ | Plastoquinone |
| $P_2$ | The general term used in this chapter for the primary electron-acceptor species |
| $(Fe)_2$ | A two-iron cluster, iron-sulfur protein presumed to be present in photosynthetic bacteria because of its esr properties at low temperature |
| $Fe_{NH\ I}$ | A nonheme iron protein (synonomous with the terminology iron-sulfur protein but less specific) assumed to be an electron carier near the primary electron acceptor of system I in oxygen-evolving organisms |
| $Fe_{NH\ II}$ | A component similar to that defined above, but viewed as interacting with the primary electron-acceptor species of system II in oxygen-evolving organisms |
| cyt | Cytochrome; a family of electron-transport proteins containing an iron-porphyrin prosthetic group (heme). These molecules undergo a reversible Fe (II)-Fe(III) valence change |
| cyt $c_2$ | A specific c-type cytochrome found in photosynthetic bacteria like *R. rubrum* |
| FMN | Flavin mononucleotide or riboflavin monophosphate |
| Car | Carotenoid |
| $H_2TTP$ | Tetraphenylporphyrin |
| BQ | Benzoquinone |
| $NADP^+$ | Nicotinamide adenine dinucleotide phosphate |

## 7.2  Bacteria, Membrane, and Isolated Fractions

| | |
|---|---|
| *R. rubrum* | *Rhodospirillum rubrum,* a photosynthetic bacterium |
| *R. spheroides* | *Rhodopseudomonas spheroides,* a photosynthetic bacterium |
| *R. capsulata* | *Rhodopseudomonas capsulata,* a photosynthetic bacterium |

| | |
|---|---|
| *Chromatium* | A photosynthetic bacterium |
| Chromatophores | Membrane vesicles prepared from photosynthetic bacteria |
| Antenna | A pigment complex that does not undergo electron transport. This complex funnels absorbed light energy to the phototrap in the photosynthetic unit |
| Phototrap | A photochemically active pigment-protein complex that contains the minimal units necessary for the primary photochemical reactions of photosynthesis. The structure is devoid of an antenna complex |
| RC | Reaction center; as used herein, synonomous with phototrap |
| PRS | Photoreceptor subunit; a phototrap that has still bound to it a functional antenna complex. It is presumed to have been incorporated into the photosynthetic membrane as a structural and functional unit |
| AUT | A procedure resulting in dissolution of photosynthetic membrane fragments by joint action of alkali, urea, and Triton x-100 (a nonionic detergent) |
| AUT-e | Purified phototrap particle obtained through the electrophoresis of material exposed to the AUT procedure. These particles often still contain much of the antenna-pigment complex |
| System I | One of what is usually assumed to be two different phototraps in oxygen-evolving organisms. The properties of this system are often compared to the phototrap of bacterial photosynthesis |
| System II | Assumed to be the second phototrap in oxygen-evolving organisms. This is the one more intimately involved in oxygen evolution |
| R-26 | A carotenoidless mutant of *R. spheroides* |

## 7.3 Miscellaneous

| | |
|---|---|
| $E_0'$ | Oxidation-reduction midpoint potential at pH 7. The value is usually understood to be relative to |

|  | a standard hydrogen electrode |
| --- | --- |
| $E_h$ | Measured oxidation-reduction potential of a particular cell relative to the standard hydrogen electrode |
| $E_{1/2}$ | Polarographic half-wave potential. Not usually given as relative to the standard hydrogen electrode because of frequent determination in aprotic solvents |
| SHE | Standard hydrogen electrode |
| SCE | Saturated calomel electrode |
| EDTA | Ethylene diamine tetraacetic acid |
| esr | Electron spin resonance |
| endor | Electron nuclear double resonance |
| $h\nu$ | One quantum of light energy |
| $\lambda_{max}$ | Wavelength of maximal absorbance for a parti-particular band in the absorbance spectrum |
| LDAO | Lauryl dimethyl amine oxide, a nonionic detergent |
| $\Phi$ | Quantum yield |
| Quantum yield | An efficiency ratio; e.g., the number of $Fe^{2+}$ cyt $c_2$ molecules oxidized per quantum of light absorbed |
| $P_i$ | Phosphate |
| ATP | Adenosine triphosphate |
| ADP | Adenosine diphosphate |
| P-lipid | Phospholipid |
| SDS | Sodium dodecyl sulfate |
| PAGE | Polyacrylamide gel electrophoresis |
| Tris | Tris(hydroxymethyl)aminomethane or 2-amino-2(hydroxymethyl)-1,3-propanediol; usually used as a buffer |
| TX-100 | Triton X-100, a nonionic detergent |

## ACKNOWLEDGMENTS

I wish to express my sincere thanks to the National Science Foundation (Grant No. GB-42860) and the National Institutes of Health (Grant No. GM 11741 and Research Career Development Award No. 5K04 GM-70133) for their support of my research over the years, which has nurtured substantial growth in my knowledge of photosynthesis. I would also like to express my appreciation to past and present members of our research group (Mr. J. Anton, Mrs. T. Cotton, Dr. B. J. Hales, Dr. R. L. Hall, Dr. M. Chu Kung, Mr. L. Morrison, Mr. D. Sponholtz, Mrs. S. Tonn, Dr. J. Runquist, and Mrs. K. Wong) who were extremely helpful in their suggestions, ideas, and constructive criticism of material for this chapter. Finally, I would like to acknowledge the enthusiastic and expert assistance of Miss Beverly Becker and Miss Madeleine Ziebka, who typed and helped organize the many drafts of this somewhat lengthy chapter.

## SUGGESTED READING

### Useful Books

R. P. F. Gregory, *Biochemistry of Photosynthesis*, Wiley-Interscience, New York, 1971.

David W. Krogmann, *The Biochemistry of Gree Plants*, Prentice-Hall, Englewood Cliffs, N. J. 1973.

R. K. Clayton, *Light and Living Matter.*, Vols. 1 and 2, McGraw-Hill, New York, 1971.

J. Lascelles, Ed., *Microbial Photosynthesis*, Dowden, Hutchinson, and Ross, Stroudsburg, Pa., 1973.

### Recent Reviews

K. Sauer, "Primary Events and the Trapping of Energy," *Bioenergetics of Photosynthesis*, Govindjee, Ed., Academic Press, Ch 3, p 115 (1975).

W. W. Parson and R. J. Cogdell, "The Primary Photochemical Reaction of Bacterial Photosynthesis", Biochim. Biophys. Acta **416**, 105 (1975).

J. T. Warden and J. R. Bolton, "Light-Induced Paramagnetism in Photosynthetic Systems," *Acc. Chem. Res.*, 7, 189 (1974).

W. W. Parson, "Bacterial Photosynthesis," *Ann. Rev. Microbiol.*, **28**, 41 (1974).

R. K. Clayton, "Primary Processes in Bacterial Photosynthesis," *Ann. Rev. Biophys. Bioeng.*, 2, 131 (1973).

J. J. Katz and J. R. Norris, Jr., "Chlorophyll and Light Energy Transduction in Photosynthesis," *Curr. Top. Bioenerg.*, 5, 41 (1973).

J. J. Katz, "Chlorophyll," *Inorganic Biochemistry*, Vol. 2, G. I. Eichhorn, Ed., Elsevier, 1973, p. 1022.

D. H. Kohl, "Photosynthesis," in *Biological Applications of Electron Spin Resonance*, H. M. Swartz, J. R. Bolton, and D. C. Borg, Eds., Wiley-Interscience, New York, 1972 pp. 213-262.

P. A. Loach and B. J. Hales, "Free Radicals in Photosynthesis," in *Free Radicals in Biology*, W. A. Pryor, Ed., Academic Press, New York, Ch 5, p. 199 (1976).

## REFERENCES

1. C. B. VanNiel, "Photosynthesis of Bacteria," in *Contributions to Marine Biology*, Stanford Univ. Press, Palo Alto, 1930, p. 158.
2. C. B. Van Niel, *Arch. Mikrobiol.*, 3, 1 (1931).
3. C. B. Van Niel, *Cold Spring Harbor Symp. Quant. Biol.*, 3, 138 (1935).
4. C. B. Van Niel, *Adv. Enzymol.*, 1, 263 (1941).
5. C. B. Van Niel, *Bact. Rev.*, 8, 1 (1944).
6. D. L. Keister and N. J. Yike, *Arch. Biochem. Biophys.*, 121, 415 (1967).
7. T. W. Englemann, *Pflügers Arch. Ges. Physiol.*, 30, 95 (1883).
8. T. W. Englemann, *Bot. Ztg.*, 46, 661-669, 667-689, 693-701, 709-720 (1888).
9. Z. Bay and R. M. Pearlstein, *Proc. Natl. Acad. Sci. U.S.*, 50, 1071 (1963).
10. G. W. Robinson, in "Energy Conversion by the Photosynthetic Apparatus," *Brookhaven Symp. Biol.*, No. 196, 16 (1967).
11. L. N. M. Duysens, *Progr. Biophys.*, 14, 1 (1964).
12. A. Yu. Borisov, *Biofizika*, 12, 630 (1967).
13. G. R. Seely, *J. Theor. Biol.*, 40, 173 (1973).
14. G. R. Seely, *J. Theor. Biol.*, 40, 189 (1973).
15. T. L. Netzel, P. M. Rentzepis, and J. Leigh, *Science*, 183, 238 (1973).
16. A. B. Rubin and L. K. Osnitskaya, *Microbiologiya*, 32, 200 (1963).
17. A. Yu Borizov and V. I. Godik, *Biochim. Biophys. Acta*, 225, 441 (1970).
18. Govindjee, J. H. Hammond, and H. Merkelo, *Biophys. J.*, 12, 809 (1972).
19. G. R. Seely in *The Chlorophylls*, L. P. Vernon and G. R. Seely, Eds., Academic Press New York, 1966, p. 523.
20. D. DeVault, in *Rapid Mixing and Sampling Techniques in Biochemistry*, B. Chance, R. H. Eisenhardt, Q. H. Gibson, and K. K. Lonbergholm, Academic Press, New York, 1964 p. 165.
21. W. W. Parson, *Biochim. Biophys. Acta*, 131, 154 (1967).
22. G. D. Case, W. W. Parson, and J. P. Thornber, *Biochim. Biophys. Acta*, 223, 122 (1970).
23. R. K. Clayton, E. Z. Szuts, and H. Fleming, *Biophys. J.*, 12, 64 (1972).
24. P. Loach, M. Chu-Kung, and B. J. Hales, *Ann. N.Y. Acad. Sci.*, 224, 297 (1974).
25. P. A. Loach and D. L. Sekura, *Biochemistry*, 7, 2642 (1968).
26. M. Baltscheffsky, *Nature*, 216, 241 (1967).
27. L. Von Stedingk and H. Baltscheffsky, *Arch. Biochem. Biophys.*, 117, 400 (1966).
28. J. B. Jackson and A. R. Crofts, *Eur. J. Biochem.*, 18, 120 (1971).
29. B. A. Melandri, A. Baccarini-Melandri, A. San Pietro, and H. Gest, *Science*, 174, 514 (1971).
30. E. Racker, in *Mechanisms in Bioenergetics*, A. San Pietro, Ed., Academic Press, New York, 1965.
31. A. W. Frenkel, *J. Biol. Chem.*, 222, 823 (1956).

32.  A. W. Frenkel, *J. Am. Chem. Soc.*, **80**, 3479 (1958).

33.  S. J. Singer, in *Structure and Function of Biological Membranes*, L. I. Rothfield, Ed., Academic Press, New York, 1971, p. 145.

34.  S. J. Singer and G. L. Nicolson, *Science*, **175**, 720 (1972).

34a.  P. A. Loach and B. J. Hales, in "Free Radicals in Biology and Pathology", Ch. 5, W. A. Pryor, Ed., Academic Press (1976), p. 199.

35  L. N. M. Duysens, Ph.D. Thesis, Utrecht, 1952.

36.  B. Kok, *Biochim. Biophys. Acta*, **22**, 399 (1956).

37.  W. Arnold and R. K. Clayton, *Proc. Natl. Acad. Sci. U.S.*, **46**, 769 (1960).

38.  I. D. Kuntz, Jr., P. A. Loach, and M. Calvin, *Biophys. J.*, **4**, 227 (1964).

39.  B. Ke, R. W. Treharne, and C. McKibben, *Rev. Sci. Instrum.*, **35**, 296 (1964).

40.  P. A. Loach and R. J. Loyd, *Anal. Chem.*, **38**, 1709 (1966).

41.  H. T. Witt and R. Moraw, *Z. Phys. Chim.*, *Neue Folge*, **20**, 253 (1959).

42.  H. T. Witt and R. Moraw, *Z. Phys. Chem.*, *Neue Folge*, **20**, 283 (1959).

43.  H. T. Witt, R. Moraw, and A. Müller, *Z. Phys. Chem.*, *Neue Folge*, **20**, 193 (1959).

44.  R. G. W. Norrish and G. Porter, *Nature*, **164**, 658 (1949).

45.  G. Porter, *Proc. R. Soc. Lond.*, **A200**, 284 (1950).

46.  G. Porter, in *Techniques of Organic Chemistry*, Vol. 8, pt. 2, A. Weissberger, Ed., Interscience, New York, p. 1055.

47.  P. A. Loach, *Biochemistry*, **5**, 592 (1966).

48.  J. C. Goedheer, *Brookhaven Symp. Biol.*, **11**, 325 (1959).

49.  B. Kok, *Biochim. Biophys. Acta*, **48**, 527 (1961).

50.  B. Commoner, J. J. Heise, and J. Townsend, *Proc. Natl. Acad. Sci. U.S.*, **42**, 710 (1956).

51.  P. Sogo, N. G. Pon, and M. Calvin, *Proc. Natl. Acad. Sci. U.S.*, **43**, 387 (1957).

52.  G. Feher, *Phys. Rev.*, **103**, 834 (1956).

53.  J. R. Norris, M. E. Druyan, and J. J. Katz, *J. Am. Chem. Soc.*, **95**, 1680 (1973).

54.  G. Feher, A. J. Hoff, R. A. Isaacson, and J. D. McElroy, *Abstr. Biophys. Soc.*, **17**, 61a (1973).

55.  P. A. Loach, G. M. Androes, A. F. Maksim, and M. Calvin, *Photochem. Photobiol.*, **2**, 443 (1963).

56.  J. H. Fuhrop and D. Mauzerall, *J. Am. Chem. Soc.*, **90**, 3875 (1968).

57.  J. H. Fuhrop and D. Mauzerall, *J. Am. Chem. Soc.*, **91**, 4174 (1969).

58.  R. H. Felton, D. Dolphin, D. C. Borg, and J. Fajer, *J. Am. Chem. Soc.*, **91**, 196 (1969).

59.  J. Fajer, D. C. Borg, A. Forman, D. Dolphin, and R. H. Felton, *J. Am. Chem. Soc.*, **92**, 3451 (1970).

60.  D. C. Borg, J. Fajer, R. H. Felton, and D. Dolphin, *Proc. Natl. Acad. Sci. U.S.*, **67**, 813 (1970).

61.  E. Rabinowitch and J. Weiss, *Proc. R. Soc. Lond.*, **162A**, 251 (1937).

62.  W. F. Watson, *J. Am. Chem. Soc.*, **75**, 2522 (1953).

63. J. J. Katz, *Dev. Appl. Spectrosc.*, **6**, 201 (1968).

64. J. J. Katz, R. C. Dougherty and L. Boucher, in *The Clorophylls*, L. P. Vernon and G. R. Seely, Eds., Academic Press, New York, 1966, p. 186.

65. G. L. Closs, J. J. Katz, F. C. Pennington, M. R. Thomas, and H. H. Strain, *J. Am. Chem. Soc.*, **85**, 3809 (1963).

66. K. Ballschmitter, K. Truesdell, and J. J. Katz, *Biochim. Biophys. Acta*, **184**, 604 (1969).

67. K. Ballschmitter, T. M. Cotton, H. H. Strain, and J. J. Katz, *Biochim. Biophys. Acta*, **180**, 347 (1969).

68. T. M. Cotton, K. Ballschmitter, and J. J. Katz, *J. Chromatogr. Sci.*, **8**, 546 (1970).

69. K. Ballschmitter and J. J. Katz, *J. Am. Chem. Soc.*, **91**, 2661 (1969).

70. J. J. Katz, K. Ballschmitter, M. Carcia-Morin, H. H. Strain, and R. A. Uphaus, *Proc. Natl. Acad. Sci. U.S.*, **60**, 100 (1968).

71. J. R. Norris, R. A. Uphaus, T. M. Cotton, and J. J. Katz, *Biochim. Biophys. Acta*, **223**, 446 (1970).

72. J. J. Katz and J. R. Norris, Jr., *Curr. Top. Bioenerg*, **5**, 41 (1973).

73. J. J. Katz, in *Inorganic Biochemistry*, Vol. II, G. I. Eichhorn, Ed., Elsevier, New York, 1973, p. 1022.

74. A. Wolberg and J. Manassen, *J. Am. Chem. Soc.*, **92**, 2982 (1970).

74a. C. E. Moore, *Atomic Energy Levels*, Vol. II, National Bureau of Standards, U.S. Government Printing Office, Washinton D.C., 1952.

75. D. Mauzerall, *Ann. N. Y. Acad. Sci.*, **206**, 483 (1973).

76. D. Mauzerall and A. Chivvis, *J. Theor. Biol.*, **42**, 387 (1973).

77. J. J. Katz, H. Strain, W. Svec, M. Studier, A. Harkness, J. Janson, B. Cope, R. Gomez-Revilla, and D. Dolphin, *J. Am. Chem. Soc.*, **94**, 7938 (1972).

78. P. A. Loach and D. L. Sekura, *Photochem. Photobiol.*, **6**, 381 (1967).

79. J. D. McElroy, G. Feher and D. Mauzerall, *Biochim. Biophys. Acta*, **172**, 180 (1969).

80. J. D. McElroy, D. C. Mauzerall, and G. Feher, *Biochim. Biophys. Acta*, **333**, 261 (1974).

81. D. H. Kohl, J. Townsend, B. Commoner, H. L. Crespi, R. C. Dougherty, and J. J. Katz, *Nature*, **206**, 1105 (1965).

82. J. R. Norris, R. A. Uphaus, and J. J. Katz, *Biochim. Biophys. Acta*, **275**, 161 (1972).

83. M. E. Druyan, J. R. Norris, and J. J. Katz, *J. Am. Chem. Soc.*, **95**, 1682 (1973).

84. G. M. Androes, M. F. Singleton, and M. Calvin, *Proc. Natl. Acad. Sci. U.S.*, **48**, 1022 (1962).

85. J. M. Olson, *Science*, **135**, 101 (1962).

86. W. J. Vredenberg and L. N. M. Duysens, *Biochim. Biophys. Acta*, **79**, 456 (1964).

87. P. A. Loach and K. Walsh, *Biochemistry*, **8**, 1908 (1969).

88. J. R. Bolton, R. K. Clayton, and D. W. Reed, *Photochem. Photobiol.*, **9**, 209 (1969).

89. C. A. Wraight and R. K. Clayton, *Biochim. Biophys. Acta*, **333**, 246

(1974).

90. R. K. Clayton, *Photochem. Photobiol.*, 1, 201 (1962).

91. R. K. Clayton, *Photochem. Photobiol.*, 5, 669 (1966).

92. S. C. Straley and R. K. Clayton, *Biochem. Biophys. Acta*, 292, 685 (1973).

93. D. W. Reed and R. K. Clayton, *Biochem. Biophys. Res. Commun.*, 30, 471 (1968).

94. R. K. Clayton and R. T. Wang, in *Methods in Enzymology*, Vol. 23, S. P. Colowick and N. Kaplan, Eds., Academic Press, New York, (1971).

95. G. Feher, *Photochem. Photobiol.*, 14, 373 (1971).

96. R. K. Clayton and R. Haselkorn, *J. Mol. Biol.*, 68, 97 (1972).

97. M. Y. Okamura, L. A. Steiner, and G. Feher, *Biochemistry*, 13, 1394 (1974).

98. L. A. Steiner, M. Y. Okamura, A. D. Lopes, E. Moskowitz, and G. Feher, *Biochemistry*, 13, 1403 (1974).

99. S. C. Straley, W. W. Parson, D. C. Mauzerall, and R. K. Clayton, *Biochim. Biophys. Acta*, 305, 597 (1973).

100. T. Beugeling *Proc. Int. Congr. Photosyn. Res. 2nd, Stresa, 1971,* **1972,** 1453.

101. T. Beugeling, L. Slooten, and P. G. M. M. Barelds-Van De Beck, *Biochim. Biophys. Acta*, 283, 328 (1972).

102. K. Sauer, E. A. Dratz, and L. Coyne, *Proc. Natl. Acad. Sci., U. S.*, 61, 17 (1968).

103. K. D. Philipson and K. Sauer, *Biochemistry*, 11, 1880 (1972).

104. J. M. Olson, K. D. Philipson, and K. Sauer, *Biochim. Biophys. Acta*, 292, 206 (1973).

105. A. Forman, D. C. Borg, R. H. Felton, and J. Fajer, *J. Am. Chem. Soc.*, 93, 2790 (1971).

106. J. R. Norris, R. A. Uphaus, H. L. Crespi, and J. J. Katz, *Proc. Natl. Acad. Sci. U.S.,* 68, 625 (1971).

107. M. W. Hanna, A. D. McLachlan, H. H. Dearman, and H. M. McConnell, *J. Chem. Phys.*, 37, 361 (1962).

108. G. Vincow and P. M. Johnson, *J. Chem. Phys.*, 39, 1143 (1963).

109. M. Chu Kung, B. J. Hales, and P. A. Loach, *Am. Soc. Photobiol., Abstr.,* **ThAM-D6,** 1973.

110. J. J. Katz, personal communication.

111. J. D. McElroy, G. Feher, and D. C. Mauzerall, *Biochim. Biophys. Acta*, 267, 363 (1972).

112. P. L. Dutton, J. S. Leigh, and M. Seibert, *Biochem. Biophys. Res. Commun.*, 46, 406 (1972).

113. J. S. Leigh and P. L. Dutton, *Biochem. Biophys. Res. Commun.*, 46, 414 (1972).

114. A. A. Krasnovskii, *Compt. Rend. (Dokl.) Acad. Sci. U.S.S.R.*, 60, 421 (1948).

115. E. I. Rabinowitch, in *Photosynthesis,*Vol II, Part 2, Interscience, New York 1956, p. 1487.

116. A. A. Krasnovskii and K. K. Voinovskaya, *Dokl. Akad. Nauk S.S.S.R.*, **66**, 663 (1949).
117. D. C. Makherjee, D. H. Cho, and G. Tollin, *Photochem. Photobiol.*, 9, 273 (1969).
118. R. A. White and G. Tollin, *Photochem. Photobiol.*, 14, 43 (1971).
119. J. Harbour and G. Tollin, *Photochem. Photobiol.*, 19, 163 (1974).
120. F. P. Schwarz, M. Gouterman, Z. Muljiani, and D. H. Dolphin, *Bioinorg. Chem.*, 2, 1 (1972).
121. R. G. Little, J. A. Anton, P. A. Loach, and J. A. Ibers, *J. Heterocyclic Chem.*, 12, 343 (1975).
122. J. A. Anton and P. A. Loach, *J. Heterocycl. Chem.*, 12, 573 (1975).
123. J. A. Anton and P. A. Loach, in preparation.
124. J. G. Komen, *Biochim. Biophys. Acta*, 22, 9 (1956).
125. C. Bril, *Biochim. Biophys. Acta*, 29, 458 (1958).
126. C. Bril, *Biochim. Biophys. Acta*, 39, 296 (1960).
127. B. Kok, *Biochim. Biophys. Acta*, 48, 527 (1961).
128. N. K. Boardman and J. M. Anderson, *Nature*, 203, 166 (1964).
129. J. M. Anderson and N. K. Boardman, *Biochim. Biophys. Acta*, 112, 403 (1966).
130. A. Garcia, L. P. Vernon, and H. Mollenhauer, *Biochemistry*, 5, 2399 (1966).
131. A. Garcia, L. P. Vernon, and H. Mollenhauer, *Biochemistry*, 5, 2408 (1966).
132. M. Y. Okamura, L. A. Steiner, and G. Feher, *Biochemistry*, 13, 1394 (1974).
133. R. L. Hall, M. Chu-Kung, M. Fu, B. J. Hales, and P. A. Loach, *Photochem. Photobiol.*, 18, 505 (1973).
134. S. Tonn, M. Chu-Kung, and P. A. Loach, 2nd Annual Meeting of the Society for Photobiology, July 22-26, 1974, Abstr. WAM-C6.
135. S. Tonn and P. Loach, manuscript in preparation.
136. P. A. Loach, R. M. Hadsell, D. L. Sekura, and A. Stemer, *Biochemistry*, 9, 3127 (1970).
137. P. A. Loach, D. L. Sekura, R. M. Hadsell, and A. Stemer, *Biochem.*, 9, 724 (1970).
138. P. A. Loach and R. L. Hall, *Proc. Natl. Acad. Sci., U.S.*, 69, 786 (1972).
139. M. Chu-Kung, Ph.D. Thesis, Northwestern Univ., 1974.
140. J. A. Culbert-Runquist and P. A. Loach, manuscript in preparation.
141. K. Weber and M. Osborn, *J. Biol. Chem.*, 244, 4406 (1969).
142. H. A. Harbury and R. H. L. Marks, *Inorg. Biochem.*, 2, 902 (1973).
143. E. Margoliash, personal communication.
144. H. Noel, M. van der Rest, and G. Gingras, *Biochim. Biophys. Acta*, 275, 219 (1972).
145. M. van der Rest, and G. Gingras, *J. Biol. Chem.*, 249, 6446 (1974).
146. M. van der Rest, H. Noel, and G. Gingras, *Arch. Biochem. Biophys.*, 164, 285 (1974).
147. L. Slooten, *Biochim. Biophys. Acta*, 256, 452 (1972).

148. L. Slooten, *Biochim. Biophys. Acta*, **314**, 15 (1973).
149. K. F. Nieth and G. Drews, *Arch. Microbiol.*, **96**, 161 (1974).
150. G. Jolchine and F. Reiss-Husson, *Fed. Eur. Biochem. Soc. Lett.*, **40**, 5 (1974).
151. G. Feher, A. J. Hoff, R. A. Isaacson, and L. C. Ackerson, *Ann. N.Y. Acad. Sci.*, **244**, 239 (1975).
152. T. M. Cotton, A. D. Trifunac, K. Ballschmitter, and J. J. Katz, *Biochim. Biophys. Acta*, **368**, 181 (1974).
153. G. Jolchine and F. Reiss-Husson, *Biochem. Biophys. Res. Commun.*, **48**, 333 (1972).
154. R. K. Clayton and B. J. Clayton, *Biochim. Biophys. Acta*, **283**, 492 (1972).
155. L. Smith, H. C. Davies, M. Reichlin, and E. Margoliash, *J. Biol. Chem.*, **248**, 237 (1973).
156. J. Aagaard and W. R. Sistrom, *Photochem. Photobiol.*, **15**, 209 (1972).
157. B. J. Segen and K. D. Gibson, *J. Bacteriol.*, **105**, 701 (1971).
158. J. Lascelles, and D. Wertlieb, *Biochim. Biophys. Acta*, **226**, 328 (1971).
159. A. E. Brown, F. A. Eiserling, and J. Lascelles, *Plant Physiol.*, **50**, 743 (1972).
160. J. L. Connelly, O. T. G. Jones, V. A. Saunders, and D. W. Yates, *Biochim. Biophys. Acta*, **292**, 644 (1973).
161. V. A. Saunders, and O. T. G. Jones, *Biochim. Biophys. Acta*, **333**, 439 (1974).
162. T. Wittenberg and W. R. Sistrom, *J. Bacteriol.*, **106**, 732 (1971).
163. W. R. Sistrom and R. K. Clayton, *Biochim. Biophys. Acta*, **88**, 61 (1964).
164. G. Drews and J. Schick, *Z. Naturforsch*, **21b**, 1097 (1966).
165. J. Schick and G. Drews, *Biochim. Biophys. Acta*, **183**, 215 (1969).
166. J. Oelze, J. Schroeder, and G. Drews, *J. Bacteriol.* **101**, 669 (1970).
167. R. L. Uffen, C. Sybesma, and R. S. Wolfe, *J. Bacteriol.*, **108**, 1348 (1971).
168. R. L. Uffen, *J. Bacteriol.*, **116**, *874 (1973)*.
169. P. Weaver, *Proc. Natl. Acad. Sci. U.S.*, **68**, 136 (1974).
170. S. Lien, A. San Pietro, and H. Gest, *Proc. Natl. Acad. Sci. U.S.*, **66**, 1912 (1971).
171. B. Marrs, and H. Gest, *J. Bacteriol.*, **114**, 1045 (1973).
172. E. Ho, K. Y. Lee, and W. R. Richards, Annual Meeting of American Society for Photobiology, 2nd, July 22-26, 1974, Abstr. WAM-C7.
173. W. W. Parson, *Ann. Rev. Microbiol.*, **28**, 41 (1974).
174. P. J. Fraker and S. Kaplan, *J. Bacteriol.*, **108**, 465 (1974).
175. P. J. Fraker and S. Kaplan, *H. Biol. Chem.*, **247**, 2732 (1972).
176. J. W. Huang and S. Kaplan, *Biochim. Biophys. Acta*, **307**, 301 (1973).
177. J. W. Huang and S. Kaplan, *Biochim. Biophys. Acta*, **307**, 317 (1973).
178. J. W. Huang and S. Kaplan, *Biochim. Biophys. Acta*, **307**, 332 (1973).
179. R. E. Hurlbert, J. R. Golecki, and G. Drews, *Arch. Microbiol.*, **101**, 169 (1974).
180. A. Gorchein, A. Neuberger, F. R. S. Tait, and G. H. Tait, *Proc. R. Soc. Lond.*, **B170**, 229 (1968).

181. A. Gorchein, *Proc. R. Soc. Lond.* **B170**, 242 (1968).

182. A. Gorchein, *Proc. R. Soc. Lond.*, **B170**, 255 (1968).

183. U. Schwenker, M. St-Onge, and G. Gingras, *Biochim. Biophys. Acta*, **351**, 246 (1974).

184. J. McElroy, G. Feher, and D. Mauzerall, Biophysical Society Meeting, 14th Feb. 25-27, 1970, Baltimore, Md., Abstr. FAM-E7.

185. G. Feher, R. A. Isaacson, J. D. McElroy, L. C. Ackerson, and M. Y. Okamura, *Biochim. Biophys. Acta*, **368**, 135 (1974).

186. P. L. Dutton and J. S. Leigh, *Biochim. Biophys. Acta*, **314**, 178 (1973).

187. J. C. Rabinowitz in *Bioinorganic Chemistry*, R. F. Gould, Ed., *Am. Chem. Soc. Publ.*, **1971**, 322.

188. W. H. Orme-Johnson in *Inorganic Biochemistry*, Vol. 2, Ed., G. I. Eichorn, Elsevier, 1973, p. 710.

189. See *Iron-Sulfur Proteins*, Vols. 1 and 2, W. Lovenberg, Ed., Academic Press, New York, 1973.

190. R. K. Clayton and S. C. Straley, *Biophys. J.*, **12**, 1221 (1972).

191. T. Herskovitz, B. A. Averill, R. H. Holm, J. A. Ibers, W. D. Phillips, and J. F. Weiher, *Proc. Natl. Acad. Sci. U.S.*, **69**, 2437 (1972).

192. B. A. Averill, T. Herskovitz, R. H. Holm, and J. A. Ibers, *J. Am. Chem. Soc.*, **95**, 3523 (1973).

193. P. G. Debrunner, C. E. Schulz, G. Feher, and M. Y. Okamura, *Biophys. J.*, **15**, 226a (1975).

194. E. S. P. Hsi and J. R. Bolton, *Biochim. Biophys. Acta*, **347**, 126 (1974).

195. G. Feher, M. Y. Okamura, and J. D. McElroy, *Biochim. Biophys. Acta*, **267**, 222 (1972).

196. R. A. Morton, in *Biochemistry of Quinones*, R. A. Morton, Ed., Academic Press, New York, 1965.

197. C. K. Mann and K. K. Barnes, in *Electrochemical Reactions in Nonaqueous Systems*, Dekker, New York, 1970, p. 190.

198. L. Slooten, *Biochim. Biophys. Acta*, **275**, 208 (1972).

199. E. J. Land, M. Simic, and A. J. Swallow, *Biochim. Biophys. Acta*, **226**, 239 (1971).

199a. R. K. Clayton, personal communication.

200. R. K. Clayton and H. F. Yau, *Biophys. J.*, **12**, 867 (1972).

201. H. Nöel, M. van der Rest and G. Gingras, *Biochim. Biophys. Acta*, **275**, 219 (1972).

202. B. Ke, A. F. Garcia, L. P. Vernon, *Biochim. Biophys. Acta*, **292**, 226 (1973).

203. Y. D. Halsey and W. W. Parson, *Biochim Biophys. Acta*, **347**, 404 (1974).

204. K. Takamiya and A. Takamiya, *Biochim. Biophys. Acta*, **205**, 72 (1970).

205. R. J. Cogdell, R. C. Prince, A. R. Crofts, *FEBS Lett.*, **35**, 204 (1973).

206. R. J. Cogdell, D. C. Brune, and R. K. Clayton, *FEBS Lett.*, **45**, 344 (1974).

207. B. J. Hales and P. A. Loach, Meeting of American Society Photobiology, 1st June 10-14, 1973, Abstr. ThPM-D5.

208. H. Rudney and W. W. Parson, *J. Biol. Chem.*, **238**, PC 3137 (1963).

209. W. W. Parson and H. Rudney, *J. Biol. Chem.*, **240**, 1855 (1965).

210. W. W. Parson and H. Rudney, *Proc. Natl. Acad. Sci. U.S.*, **53**, 599 (1965).

211. L. Morrison, J. A. Culbert-Runquist, and P. A. Loach, manuscript in preparation.

212. S. J. Tonn, Thesis. Northwestern University (1976).

213. P. A. Loach, R. A. Bambara, and F. J. Ryan, *Photochem. Photobiol.*, **13**, 247 (1971).

214. J. Fajer, D. C. Borg, A. Forman, R. H. Felton, D. Dolphin, and L. Vegh, Brookhaven Nuclear Laboratory Report 18246.

215. J. Fajer, D. C. Borg, A. Forman, D. Dolphin, and R. H. Felton, *J. Am. Chem. Soc.*, **95**, 2739 (1973).

216. R. H. Felton and H. Linschitz, *J. Am. Chem. Soc.*, **88**, 1113 (1966).

217. P. Loach, unpublished results.

218. L. N. M. Duysens, *Brookhaven Symp. Biol.*, **11**, 18 (1958).

219. R. T. Ross and M. Calvin, *Biophys. J.*, **7**, 595 (1967).

220. R. S. Knox, *Biophys. J.*, **9**, 1351 (1969).

221. G. Doring, H. H. Stiehl, and H. T. Witt, *Z. Naturforsch.*, **22b**, 639 (1967).

222. G. Doring and H. T. Witt, *Proc Intern. Congr. Photosynth. Res. 2nd, Stresa, Italy, June 24-29*, (1971), 39.

223. R. Malkin and A. J. Bearden, *Proc. Natl. Acad. Sci. U.S.*, **68**, 16 (1971).

224. A. J. Bearden and R. Malkin, *Biochem. Biophys. Res. Commun.*, **46**, 1299 (1972).

225. J. Amesz, M. P. J. Pulles, and B. R. Velthuys, *Biochim. Biophys. Acta*, **325**, 472 (1973).

226. T. Cotton, J. J. Katz, and P. A. Loach, unpublished results.

227. F. Fong, *Proc. Natl. Acad. Sci. U.S.*, **71**, 3692 (1974).

228. B. R. Masters and D. Mauzerall, Annual Biophysical Society Meeting, 19th, Feb. 18-12, 1975, Abstr. F-AM-12.

229. N. K. Bridge and G. Porter, *Proc. R. Soc. Lond.*, **A244**, 276 (1958).

230. G. E. Adams and B. D. Michael, *Trans, Faraday Soc.*, **63**, 1171 (1967).

231. J. R. Norris, H. Scheer, M. E. Druyan, and J. J. Katz, *Proc. Natl. Acad. Sci. U.S.*, **71**, 4897 (1974).

232. J. L. Hoard, *Science*, **174**, 1295 (1971).

233. R. A. Uphaus, J. R. Norris, and J. J. Katz, *Biochem. Biophys. Res. Commun.*, **61**, 1057 (1974).

234. H. A. Harbury, *J. Biol. Chem.*, **225**, 1009 (1957).

235. P. A. Loach and J. J. Katz, *Photochem. Photobiol.*, **17**, 195 (1973).

236. J. R. Bolton and K. Cost, *Photochem. Photobiol.*, **18**, 417 (1973).

237. T. Hiyama and B. Ke, *Proc. Natl. Acad. Sci. U.S.*, **68**, 1010 (1971).

238. T. Hiyama and B. Ke, *Arch. Biochem. Biophys.*, **147**, 99 (1971).

239. B. Ke, *Biochim. Biophys. Acta*, **301**, 1 (1973).

240. M. Y. Okamura, R. A. Isaacson, and G. Feher, *Proc. Natl. Acad. Sci. U. S.*, **72**, 3491 (1975).

241. M. G. Rockley, M. W. Windsor, R. G. Cogdell, and W. W. Parson, *Proc. Natl. Acad. Sci. U. S.*, **72**, 2251 (1975).

242. P. A. Loach, *Biophys. J.*, **15**, 225a, ABS TH-PM-L10 (1975).

243. W. L. Butler, *Accounts Chem. Res.*, **6**, 177 (1973).
244. C. F. Fox and N. Tsukagoshi, in *Membrane Research*, C. F. Fox, Ed., Academic Press, New York, 1972, pp. 145-151.
245. L. Lin and J. P. Thornber, *Photochem. Photobiol.*, **22**, 37 (1975).
246. K. F. Nieth, G. Drews, and R. Feick, *Arch. Microbiol.*, **105**, 43 (1975).
247. P. A. Loach, third annual American Society for Photobiology meeting, June 22-26, 1975, ABS MPM-C4.

# A BASIS FOR BIOLOGICAL PHOSPHATE AND SULFATE TRANSFERS — TRANSITION STATE PROPERTIES OF TRANSFER SUBSTRATES

K. T. DOUGLAS

*Department of Chemistry*
*Duquesne University*
*Pittsburgh, Pennsylvania*

# 1  INTRODUCTION

The level of cellular ATP has been shown to remain fairly constant, indicating cellular control of the rate of ATP synthesis and/or the rate of ATP regeneration [1]. This is largely achieved by phosphate transfers. It is the purpose of this article to comment on the molecular mechanisms of acyl* transfer and to show how these may be

---

*The term acyl is intended to cover the carboxyl ($-CO\cdot O-$) as well as the several phosphoryl ($>PO\cdot O-$) and sulfonyl ($-SO_2\cdot O-$) groupings possible. Phosphoryl refers collectively to phosphoryl [$(RO)_3PO$], phosphonyl [$(RO)_2-PO\cdot R'$], and phosphinyl [$R_2PO\cdot OR'$] substitution patterns for P(V) phosphorus. The term sulfonyl is used similarly with respect to sulfur substitution.

related to biological systems.

The physical organic chemistry of acyl transfer (particularly at the carboxyl and phosphoryl levels) has now been brought to a degree of sophistication sufficient to permit the drawing of some fairly sound inferences. In the first part of this chapter we look at the ways in which information has been obtained on the nature of the nonenzymatic processes for acyl transfer from biological substrates. From this we go on to construct a general panorama of the activity of these species and the rules governing their reactivity. Finally, some implications of the "transition state chemistry" derived from the *in vitro* studies are described for the biochemistry of these substrates.

## 2  GROUP TRANSFER IN METABOLIC PROCESSES

A generalized phosphate donor is represented by **I**, some of the specific instances found in nature being collected in Table 1, along

### TABLE 1    SOME BIOLOGICAL ACYL DONORS DERIVED FROM I

| R | Class |
| --- | --- |
| Alkyl, aryl | Phosphate monoesters |
| AMP, ADP | ADP, ATP |
| Phosphates | Pyrophosphates |
| Guanidino | Creatine or arginine phosphates |
| Acetyl | Acetyl phosphate |
| Glycero | Glycerol phosphate |
| Enolpyruvyl | Phosphoenolpyruvate |
| $-CONH_2$ | Carbamyl phosphate |
| Sulfates | Phosphosulfates[a] |

[a] For example, PAPS, which acts as a sulfate donor by S—O cleavage.

with some other acyl donors.

$$\begin{array}{c} O \\ \parallel \\ (H)O-P-OR \\ \mid \\ O(H) \end{array}$$

**I**

Cellular energy levels are ultimately dictated by the phosphate-transferring equilibrium between ATP and ADP. Enzymes involved in this include the various kinases, for example, hexokinase and phosphotofructokinase. Such enzymes are often critically disposed in the regulation of whole metabolic routes. Thus the phosphorylase a and b systems control glycogen metabolism [2] and the pyrophosphatases (e.g., the feedback-controlled, glutamine-5′-phosphoribosyl pyrophosphate amidotransferase) direct the first reaction specifically involved in purine biosynthesis [3]. Creatine phosphate buffers the ATP level in muscle with the aid of creatine kinase [4]. The likelihood that the sodium pump is in fact sodium-potassium adenosine triphosphatase [5, 6] should also be mentioned here. Acetyl phosphate serves as an acetyl and phosphate donor [7]. Carbamyl phosphate, which, in principle, can act as either a phosphoryl or carbamyl donor, is important in the early states of purine and pyrimidine biosynthesis under the influence of aspartate transcarbamylase [8a] and in the urea cycle [8b]. 3′-Phosphoadenosine-5′-phosphosulfate (PAPS) is a biological sulfate donor to a variety of species [9].

The centrality of these highly activated group-transferring species in biochemistry is unquestionable. Furthermore, many of the steps of biosynthesis depend on a phosphate transfer from such a substrate mediated by an enzyme, which is often subject to some form of regulatory control [3].

In this essay we probe the mechanistic chemistry of a broad array of substrates, which includes those in Table 1, and show them to possess a common, dominant feature. We may write the general structure **II** for a derivative that has an acidic group (HX—) in the alpha position with respect to the acyl group ($Y = C$, $-\overset{\overset{O}{\|}}{P}R-$, $-\overset{\overset{O}{\|}}{\underset{\underset{O}{\|}}{S}}-$) and also has an active leaving group (L). The chemistry of many such

$$HX-\overset{\overset{O}{\|}}{Y}-L \qquad\qquad \overset{N\diagdown}{\phantom{.}}{\delta^- X\text{---}Y}\overset{\diagup O}{\underset{\diagdown L^{\delta-}}{}}$$

**II**                              **III**

species in group **II** is well established and the transition states for hydrolysis and acyl transfer are described as ElcB, the rate-determining step being a unimolecular elimination (E1) from the conjugate base (cB) of the substrate with little interaction with the nucleophile (N) **III**. Having defined the transition states for **I** and **II**, we continue by looking at some enzymatic reactions involving these acidic substrates and measure the impact of the physical organic studies in this area. As a detailed, comprehensive review [10] of the elimination routes leading to acyl transfer has appeared, we attempt to be constructively selective and point out the *generalized* features of these important reactions. The work discussed has, for the most part, been carried out in predominantly aqueous solution; the mechanistic situation for such substrates in nonaqueous solution has been by no means as fully investigated, although it is undoubtedly a ripe field.

## 3  MECHANISMS OF NONENZYMATIC ACYL TRANSFER FROM IONIZABLE SUBSTRATES

Whereas esters such as phenyl benzoate hydrolyze only by a bi-molecular attack of hydroxide ion on the neutral ester in basic solution, the presence of an acidic hydrogen on the atom alpha to the acyl group leads to two additional, mechanistically distinct routes: bimolecular attack of water on the conjugate base of the substrate (an addition-elimination, AE*, pathway) and unimolecular elimination from the anionic substrate (EA). These processes are summarized in Scheme 1 for the rapid preequilibrium formation of the conjugate base of the substrate. Cleavage of the H–X bond can be general-base catalyzed in certain cases (e.g., carbon-acidic substrates). However, most of the chapter is concerned with oxygen- and nitrogen-acids, for which proton transfers are diffusion controlled, so that Scheme 1 is widely applicable. The term for AE attack of hydroxide ion on the anionic substrate has been omitted from Scheme 1.

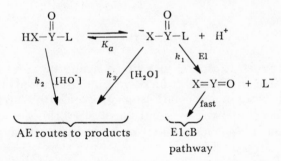

Scheme 1

*The terms addition-elimination (AE) and elimination-addition (EA) refer to processes in which *nucleophiles*, as opposed to bases, attack in the initial and later states of reaction, respectively.

The AE routes ($k_2$ and $k_3$) need be described in no greater detail at this stage than by saying that the slow step is attack of nucleophile, as excellent reviews of the bimolecular reactions of carboxylic [11, 12], phosphoric [13-16], and sulfonyl [17] derivatives are already available. It is sufficient to say that for an AE mechanism at any acyl center (eq. 1) a formally stepwise mechanism may be written. If the lifetime of the intermediate is very small, the overall process appears concerted.

$$N + \underset{L}{\overset{O}{\underset{\Vert}{Y}}} \underset{k_{-1}}{\overset{k_1}{\rightleftharpoons}} \overset{O^-}{\underset{\vert}{^+N-Y-L}} \xrightarrow{k_2} {^+N-Y}{\overset{O}{\diagup}} + L^- \qquad (1)$$

We now turn our attention to the E1cB mechanism, and after a few specific examples of the route, we look at the unified features of these reactions. Finally, we discuss the possible interactions in enzyme systems involving such E1cB substrates.

### 3.1  Examples of E1cB Acyl Transfers and Solvolyses

As early as 1935 Miller and Case [18] had formulated the alkaline hydrolysis of diethyl carbonate as a two-stage process involving a monoester anion, whose unimolecular decomposition was rate determining (**IV**). If a carbonate half-ester is involved in biotin action

$$EtO\overset{\frown}{-}C\overset{O}{\underset{\underset{O^-}{\frown}}{\diagup}}$$

**IV**

[19], this type of reaction may be of significance and is an obvious model for decarboxylase action of enzymes (discussed later).

The hydrolyses of carbamate esters proceed with E1cB formation of an isocyanate, followed by rapid solvolysis of this species [20-23] as in eq. 2. Such a scheme is noteworthy because carbamyl phosphate

$$ArNH\cdot CO\cdot OR \underset{}{\overset{HO^-}{\rightleftharpoons}} Ar\overset{-}{N}\cdot CO\cdot OR \xrightarrow{slow} ArN{=}C{=}O + RO^-$$
$$\overset{fast}{\diagdown} H_2O$$
$$\downarrow$$
$$products \qquad\qquad (2)$$

can react in a similar manner. The hydrolyses of acetoacetic esters (eq. 3) are mediated by a ketene formed by an E1cB reaction [24].

$$CH_3COCH_2CO \cdot OR \underset{BH^+}{\overset{B}{\rightleftharpoons}} CH_3CO \cdot {}^-CH \cdot COOR$$

$$CH_3CO \cdot CH=C=O \ + \ OR^- \xrightarrow[fast]{H_2O} CH_3COCH_2CO_2^- \ + \ ROH$$

(3)

In P(V) reactions such an elimination mode is also adopted by α-acidic esters; for example, phosphate monoesters [25, 26] and phosphoramidic derivatives [27-31] react via metaphosphate ion (V), and metaphosphorimidate ion (VI), respectively (see eq. 4 and 5).

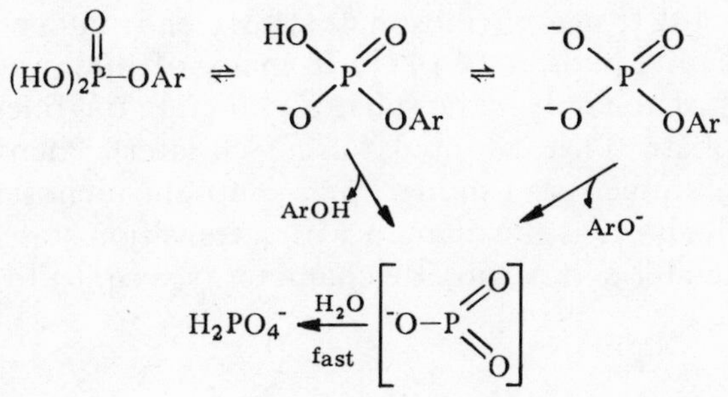

(4)

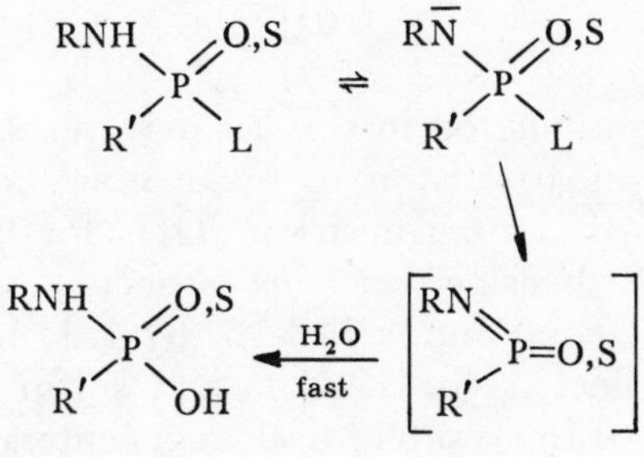

**V**

(5)

**VI**

Similarly, elimination-addition routes occur at the sulfonyl level for suitable acidic species, for example, sulfate monoesters [32, 33], sulfamic derivatives [34, 35], and aryl α-toluenesulfonates [36], as shown in eq. 6a, 6b, and 6c, respectively.

$$HXSO_2Y \; \rightleftharpoons \; ^-XSO_2Y \; \longrightarrow \; [X{=}SO_2] \; + \; Y^-$$

$$\overset{\displaystyle H_2O}{\diagup}{}^{fast}$$

$$HXSO_3H \qquad\qquad (6)$$

(a)  X = —O

(b)  X = —NR

(c)  X = —CHPh

All of the substrates in Table 1 may react according to one of these schemes. The metaphosphate mechanism (eq. 4), which has been discussed thoroughly [13-16, 37-40], occurs for phosphate mono-esters [13-16, 37-40] (but not diesters [41]), polyphosphates (including ATP, ADP, and pyrophosphate [13-16, 37-40]), creatine phosphate [42], acetyl phosphate [43], and, under certain conditions, carbamyl phosphate [44]. In many of these cases, the ionization states of the polyacidic substrates dictate the finer details of the metaphosphate route adopted (discussed later). Phenyl phosphosulfate (VII), studied as a model for 3′-phosphoadenosine 5′-phosphosulfate, solvolyzes as the dianion with a transition state that probably has considerable sulfur trioxide character (see eq. 6) [45, 46].

$$\begin{array}{ccc} O & & O \\ \| & & \| \\ PhO{-}P{-}O{-}S{-}O^- \\ {}_-| & & | \\ O & & O \end{array}$$

**VII**

Data have so accumulated in the last few years for such reactions that we are now in a position to make some confident statement of the properties of E1cB transition states for hydrolysis and acyl transfer. Evidence bearing on this aspect has been thoroughly reviewed for P(V) compounds [13-16, 37-40], but data have only recently been collected for carbon and sulfur centers [10]. No unified discussion of the results for all acyl centers has yet appeared.

## 3.2  Approaches to the Mechanism of Ionizable Ester Solvolyses

It is apposite to summarize the main approaches that have been used to illuminate the mechanistic features of E1cB reactions. In this way the extent of their applicability to enzyme systems can be gauged. These approaches are discussed in terms of the following characteristics:

1. Kinetic rate law and buffer effects
2. α-Hydrogen exchange and kinetic isotope effects
3. Intermediate and related studies
4. Rate comparisons with model compounds
5. Leaving-group investigations
6. Activation parameters
7. Stereochemical analyses

An exhaustive survey is unnecessary, as a fuller treatment is available elsewhere [47].

*3.2.1.* **Kinetic Laws and Buffer Effects.** *3.2.1.1. Diffusion-Controlled, Preequilibrium Proton Transfer.* For an ionizing substrate (HX·Y(O)L) the minimal kinetic scheme has already been given (Scheme 1). By far the largest group of substrates studied are oxygen- or nitrogen-based acids and are most appropriately discussed in this section. The kinetic expression for Scheme 1 is eq. 7, where $[HX \cdot Y(O)L]_{TOT}$ is the total substrate concentration, $K_w$ is the autoprotolysis constant of water, and $K_a$ the acid dissociation constant of the substrate, other constants being defined as in Scheme 1.

$$\frac{velocity}{[HX \cdot Y(O)L]_{TOT}} = \frac{(k_1 + k_3 + k_2 K_w / K_a)}{(1 + K_w / K_a [HO^-])} , \qquad (7)$$

or under pseudofirst order conditions,

$$k_{obs} = \frac{k'}{(1 + K_w / K_a [HO^-])} . \qquad (8)$$

As proton transfer is fast, the presence of nonlyate bases serves only to shift the ionization equilibrium of the substrate by a secondary effect on lyate ion concentration. In aqueous solution the situation is equivalent to specific-base catalysis. Rate equation 8 is algebraically analogous to the celebrated Michaelis-Menten equation of enzyme kinetics. In this case when the rate is half the limiting value ($k'$), the hydroxide ion concentration is equal to $K_w / K_a$. Thus, depending on the acid dissociation constant of the substrate ($K_a$), saturation kinetics may be observed, for example, aryl methylaminosulfonates [35] and aryl $N,N'$-diphenylphosphorodiamidates [29, 30, 48]. If the p$K_a$ of the substrate is high, only the initial, linear portion of the saturation dependence of $k_{obs}$ on hydroxide ion concentration may be observed, for example, aryl and alkyl $N$-phenylcarbamates [21].

The pH dependency of the kinetic behavior of diacids is complicated and the shape of the $k_{obs}$ versus hydroxide ion concentration plots depends on the relative reactivities of the mono- and dianions toward water and hydroxide ion. For this reason a bell-shaped pH profile is often observed, as for some aryl phosphate hydrolyses [26].

*3.2.1.2 Buffer Effects.* The bimolecular hydrolyses of activated esters usually involve considerable buffer catalysis, particularly by amines. The ElcB route for diffusion-controlled proton transfer systems would not be expected to exhibit buffer catalysis, but the absence of such rate effects can be, at best, corroborative of an elimination [21]. Further discussion of the effects of buffer and added species in E1cB hydrolyses is offered later

*3.2.1.3 Carbon-Acids: the Possibility of Rate-Determining Proton Transfer.* General bases can, in this case, catalyze deprotonation of the substrate, and saturation kinetics will be obtained for $k_{obs}$ versus free base concentration, if the scheme followed is true ElcB (eq. 9) or

$$SH \underset{[BH^+]\,k_{BH^+}}{\overset{[B]\,k_B}{\rightleftharpoons}} S^- \overset{E1}{\underset{k_1}{\longrightarrow}} \text{products} \qquad (9)$$

"blind-alley" (eq. 10).

$$^-S \underset{k_B\,[B]}{\overset{k_{BH^+}\,[BH^+]}{\rightleftharpoons}} SH \overset{HO^-}{\underset{B_{AC}2}{\longrightarrow}} \text{products} \qquad (10)$$

Pratt and Bruice [24] and Kirby and Lloyd [49] have obtained curved plots for carbon-acidic substrates in amine buffers. Hence the observation of saturation in a plot of $k_{obs}$ versus free base concentration for an acidic substrate is not of itself sufficient demonstration of an E1 process, whether from nitrogen, carbon, or other acid. Nor is the absence of detectable curvature evidence against an E1 process if the p$K_a$ of the substrate is high (or expected to be).

**3.2.2 Alpha-Hydrogen Exchange and Kinetic Isotope Effects.** For ester solvolysis following an E1cB route, the condition $k_{exch} > k_{hydrol}$ is necessary but not sufficient [50]. Casanova et al. found that $k_{exch}/k_{hydrol} = 8$ for $p\text{-Me}_2\overset{+}{S}C_6H_4\cdot CH_2CO_2Me$, which probably hydrolyzes by an ion-pair route [50]. For N-acidic esters, which hydrolyze by E1cB routes, rapid and complete alpha hydrogen/-deuterium exchange is observed, for example, for phenyl *N*-methylcarbamate [23]. When the substrates are N- or O-acids, the proton transfers are likely to be diffusion controlled and the criterion $k_{exch}$ > $k_{hydrol}$ is readily met. When preequilibrium deprotonation is not *much* faster than the overall reaction (C-acids), product composition studies after partial reaction in deuterated solvent can be diagnostic [51]. The uptake of one, and only one, deuterium in the base-promoted deuterolysis of alkane-sulfonyl chlorides has been cited as evidence of an intermediate sulfene, and a review of this field has appeared [52]. However, recent evidence of multiexchange has pointed to a more complicated set of zwitterionic reaction equilibria [53].

Tobias and Kézdy proposed that bimolecular ($B_{AC}2$) and E1cB routes could be distinguished because there should be no kinetic solvent deuterium isotope effect on $k'$ (= $k_1$) for the E1cB route, whereas for the bimolecular route of hydroxide ion attacking neutral ester ($k' = k_2 K_w/K_a$ from eq. 7 and 8) a significant effect can be expected. Although there is some precedent for this view in the literature, Table 2 shows that in general $k'_H \neq k'_D$ and solvent isotope effects are not often diagnostic.

### TABLE 2   KINETIC SOLVENT ISOTOPE EFFECTS ON E1cB REACTIONS $(k')$

| Substrate | $k_D/k_H$ | Reference |
|---|---|---|
| $^{2-}O_2P(O)OAr$ | 1.00 | 26 |
| $^-OP(OH)(O) \cdot OdNp$ (2,4) | 0.69 | 26 |
| $^-OP(OH)(O)SpNp$ | 0.56 | 54 |
| $^-OP(OH)(O)SPh$ | 0.69 | 54 |
| $Ph\overline{N}P(NHPh)(O)OpNp$ | 0.77 | 29 |
| $^-OSO_2OpNp$ | 0.79 | 32 |
| $Me\overline{N} \cdot SO_2OpNp$ | 0.74 | 35 |
| $CH_3CO \cdot \overline{C}H \cdot CO \cdot OpNp$ | 0.87 | 24 |

*3.2.3* **Intermediate and Related Studies.** The E1cB route is characterized by an ionic and an electrophilic heterocumulene intermediate. Direct observations during solvolysis of the electron-deficient species have been precluded in general by their high labilities in hydroxylic solvents, and evidence for their occurrence is often indirect. However, phenyl isothiocyanate has been detected spectrally in the hydrolysis of 4-acetylphenyl *N*-phenylthiocarbamate [55]. Attempts to trap ketenes by cycloaddition to 1,3-dienes failed [24], although successful diversions have been achieved using amines and other nucleophiles.

If the ElcB route is given by eq. 11, the attack of nucleophiles

$$
\begin{array}{c}
\overset{\text{fast}}{\underset{L}{\overset{}{\rightleftharpoons}}} X^- \xrightarrow[\underset{+L^-}{\text{slow}}]{k_4} X \xrightarrow[\text{fast}]{k_5,NH} X-H
\end{array} \tag{11}
$$

(including solvent species) is post-rate controlling, and increasing the concentration of added nucleophiles at constant pH should not affect the rate of cleavage of the substrate ($k_4$) but could change the nature of the products (from $k_5$). Hegarty and Frost showed that, although the presence of 4-chloroaniline did not change the rate of loss of 4-nitrophenolate ion from 4-nitrophenyl N-phenylcarbamate, the products were more than 98% diverted to the anilide [22]. Similar results have been obtained with 4-nitrophenyl methylaminosulfonate [35], 4-nitrophenyl N-phenyl phenylphosphonamidate [31], and 4-nitrophenyl N,N'-diphenylphosphorodiamidate [31]. In the hydrolysis of acetyl phosphate monoanion, not only is the yield of fluorophosphate increased with added fluoride ion, but also the rate of substrate disappearance; fluoride ion probably reacts nucleophilically at phosphorus [43], towards which it is extraordinarily effective (better than ethoxide or thiophenolate ions with diisopropylphosphorofluoridate [56]).

The rapid reaction of X (eq. 11) with nucleophiles is the basis of the diagnostic use of mixed solvents in this area. If X is sufficiently "hot," it may be expected to acylate the components of a suitable mixed-solvent system nonselectively [57, 58]. Nonselectivity in phosphorylation and sulfonylation has been observed, but this is the exception rather than the rule, and the lack of selectivity depends on the fortuitous choice of temperature and solvent [38].

### 3.2.4  Rate Comparisons.

Early evidence of disparity in mechanism between ionizable and fully substituted acyl compounds came from the large rate differences in the urethan series. Dittert [20] found that 4-nitrophenyl N-methylcarbamate hydrolyzes a million times as fast as 4-nitrophenyl N,N-dimethylcarbamate. The rate enhancement in the aminosulfonate series is even greater: the ratio of (apparent) second order hydroxide ion rate constants is $10^8$ for 4-nitrophenyl methylaminosulfonate and dimethylaminosulfonate [35]. There are dangers in ascribing drastic changes in mechanism on the basis of rate comparisons unless the ratio is very large. The actual ratio obtained for an E1cB process compared to the corresponding bimolecular route depends on steric factors, the leaving group, and the temperature.

*3.2.5  Leaving-Group Effects.* Unimolecular eliminations show high rate dependence on leaving group (LG) parameters (Hammett substituent constants or leaving group $pK_a$'s) and relatively abrupt changes in mechanism with change in departing group. Generally $\rho$ and $\beta_{LG}$ values* for E1cB hydrolyses are much higher than those for their bimolecular analogs. Table 3 shows that for E1cB routes $\rho$ values of 2.6 to 3.2 (and $\beta = -1.2$ to $-2.9$) are common, whereas for the corresponding bimolecular process, the range for $\rho$ is 0.85 to 1.55 (and $\beta = -0.05$ to $-0.4$). It is also typical that good correlations are obtained for E1cB processes when Hammett sigma-minus values are used. Such high $\rho$ values and sigma-minus dependence, indicating pronounced phenolate character in the transition state, are to be expected for the fragmentation scheme ($k_4$ in eq. 11).

Bimolecular ester solvolyses follow Hammett sigma relationships because the rate-determining step is attack of hydroxide ion. Although unlikely for N- and O-acids under normal conditions of diffusion-controlled deprotonation, the correlation with sigma-minus might be caused by concerted (E2) elimination. Inverse, kinetic solvent isotope effects ($k_{DO^-} > k_{HO^-}$) have been used to exclude this in carbamic ester hydrolysis [21]. Concerted general-base catalysis has been excluded as the mechanism of hydrolysis of aryl *N*-methyl-aminosulfonates using the rule, proposed by Jencks [59], whereby concerted general-base catalysis can occur only when the $pK_a$ of the base lies well within the $pK_a$ values of products and reactants (in this instance the $pK_a$ of product is probably $<0$ and that of the reactant is $\sim9$) [35].

*3.2.6  Activation Parameters.* Activation parameters can, in principle, distinguish E1cB from bimolecular mechanisms, since the former, a fragmentation, should show a considerably more positive entropy of activation than the latter [60]. However, resolution of the forming fragments in the transition state sometimes acts to decrease the positive contribution to $\Delta S^{\ddagger}$ from cleavage in aqueous solution. A selection of values for various substrates in aqueous solution is given in Table 4. In many cases $\Delta S^{\ddagger}$ values for known E1cB reactions are indeed around zero—the values for bimolecular models are considerably more negative. Care must be exercised in deducing mechanism on the basis of the activation parameters, since the $\Delta S^{\ddagger}$

---

* $\beta_{LG}$ is the Brønsted exponent for leaving group variation; it is the slope of a plot of $\log_{10}$ (rate constant) versus the $pK_a$ of the conjugate acid of the leaving group.

## TABLE 3  LEAVING-GROUP SENSITIVITIES FOR ESTER SOLVOLYSES IN BASIC SOLUTION

| Series[a] | Molecularity | $\rho$ | $-\beta_{LG}$ | Correlation | Reference |
|---|---|---|---|---|---|
| $NMe_2 \cdot CO_2Ar$ | 2 | — | -0.4 | $pK_{LG}$ | 145 |
| $NHMe \cdot CO_2Ar$ | 1 | — | -1.3 | $pK_{LG}$ | 145 |
| $NH_2 \cdot CO_2Ar$ | 1 | — | -1.3 | $pK_{LG}$ | 145 |
| $MeNPh \cdot CO \cdot OAr$ | 2 | 1.3 | $-0.6^b$ | $\sigma$ | 22 |
| $PhNH \cdot CO_2R$, Ar | 1 | 2.9 | -1.15 | $\sigma - (pK_{LG})$ | 21 |
| $CH_3COCH_2CO_2Ar$ | 1 | — | -1.29 | $pK_{LG}$ | 24 |
| $CH_3COCH_2CO_2R$ | 2 | — | -0.05 | $pK_{LG}$ | 24 |
| $(EtO)_2P(O)OAr$ | 2 | 1.26 | — | $\sigma$ | 146 |
| cyclic phosphate structure $\;2^-O_2P(O)OAr$ | 2 | 1.18 | — | $\sigma$ | 67 |
| $(PhNH)_2P(O)OAr$ | 1 | $2.6^c$ | — | $\sigma^-$ | 26 |
| $(PhNH)_2P(S)OAr$ | 1 | 2.8 | — | $\sigma^-$ | 29 |
| $(Ph\overline{N})P(O)(NHPh) \cdot OAr$ | 1 | 2.8 | — | $\sigma^-$ | 30 |
| | 1 | $3.08^c$ | — | $\sigma^-$ | 30 |

| | | | | | |
|---|---|---|---|---|---|
| (PhNH)P(O)Ph·OAr | 1 | 2.2 | — | $\sigma$ | 31 |
| (OC$_6$H$_8$N)$_2$P(O)·OAr | 2 | 1.47 | — | $\sigma$ | 29 |
| PhSO$_2$OAr | 2 | 2.75 | — | $\sigma$ | 17 |
| $^-$OSO$_2$OAr | 1 | -1 | -1.2[c] | p$K_{LG}$ | 33 |
| MeNH·SO$_2$OAr | 1 | 6.4 | -2.9[b] | $\sigma^-$, (p$K_{LG}$) | 35 |
| Me$_2$N·SO$_2$OAr | 2 | — | -1.1 | p$K_{LG}$ | 35 |
| PhCH$_2$SO$_2$OAr | 1 | 4.84 | -2,14[b] | $\sigma^-$(p$K_{LG}$) | 36 |

[a] Ar = substituted phenyl esters; R = substituted alkyl esters.

[b] Calculated using the $\rho$ for phenol ionization [144].

[c] These values refer to unimolecular rate constants and not the *overall* reaction rate constants ($k_{HO^-}$) which include an ionization term. See footnote on p. 231.

## TABLE 4   ENTROPIES OF ACTIVATION FOR SOME ESTER HYDROLYSES IN BASIC SOLUTION

| Compound | $\Delta S^{\ddagger}$ (eu at 25°C) | Reference |
|---|---|---|
| PhNH·CO$_2$Et | −25.5 | 147 |
| PhNMe·CO$_2$Et | −39.3 | 147 |
| PhNH·COOPh | + 5.0 | 147 |
| PhNMeCOOPh | −28.0 | 147 |
| MeNH·CO·O—(ring with $\overset{+}{N}Me_3$) | +11.4 | 147 |
| Me$_2$N·CO·O—(ring with $\overset{+}{N}Me_3$) | −27.0 | 147 |
| MeNH·CO·OAr | +2 to +10 | 145 |
| PhOPO$_3$H$^-$ | + 0.9 | 60 |
| α-napthyl-OPO$_3$H$^-$ | + 4.1 | 60 |
| β-naphthyl-OPO$_3$H$^-$ | + 5.5 | 16 |
| (ring with HO$_2$C)—OPO$_3$H$^-$ | − 1.4 | 16 |
| BuS·PO$_3$H$^-$ | + 3.1 | 16 |
| EtO$_2$C·NH·PO$_3$H$^-$ | +14.4 | 16 |
| PhCONH·PO$^3$H$^-$ | + 7.5 | 60 |
| AcO·PO$_3$Ph$^-$ | −29 | 60 |
| AcO·PO$_3$H$^-$ | + 3.7 | 60 |
| AcOPO$_3^{2-}$ | − 3.6 | 60 |
| NO$_2$—(ring with NO$_2$)—OPO$_3^{2-}$ | + 6.6 | 26 |
| pNpOSO$_3^-$ | −18.0 | 32 |

values obtained from the bimolecular rate constants for E1cB processes are composite ($k_{HO^-} = k'K_a/K_w$) and should therefore be dissected to allow for the preionization.

*3.2.7 Stereochemical Analysis.* If, by analogy to the isoelectronic sulfur trioxide, the postulated metaphosphorimidate ion **VIII** is planar, then, Gerrard and Hamer [28] argued, solvolysis of an optically active substrate should lead to racemization as a bimolecular displacement at phosphorus would lead to inversion. The neutral hydrolysis of **IX** (X = Cl) was stereospecific, whereas *alkaline hydrolysis* led to racemization, employing a change in mechanism to a symmetrical intermediate for the latter. When X = 4-nitrophenyl, alkyline hydrolysis of **IX** leads to products of stereochemistry inconsistent with the E1cB route [61].

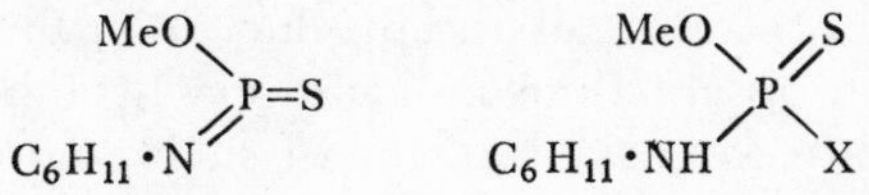

**VIII**          **IX**

## 3.3 The Nature of the E1cB Transition State

In this section the nature of the *general* ElcB transition state is derived; evidence and examples are drawn from all acyl levels, carboxyl, phosphyl, and sulfyl. The very high (linear free energy) dependences on leaving groups (Table 3) in unimolecular compared to bimolecular processes is strong evidence of advanced bond cleavage in the transition state; the rate-determining step in the bimolecular path reflects attack of the nucleophile, whereas in unimolecular cases it mirrors solely the fission of the bond restraining the departing group. Often a correlation of rates with Hammett sigma-minus constants is observed for suitable E1cB-active esters (Table 3), implying marked phenolate ion character in the transition state [10], since mesomeric stabilization (Scheme 2) is possible.

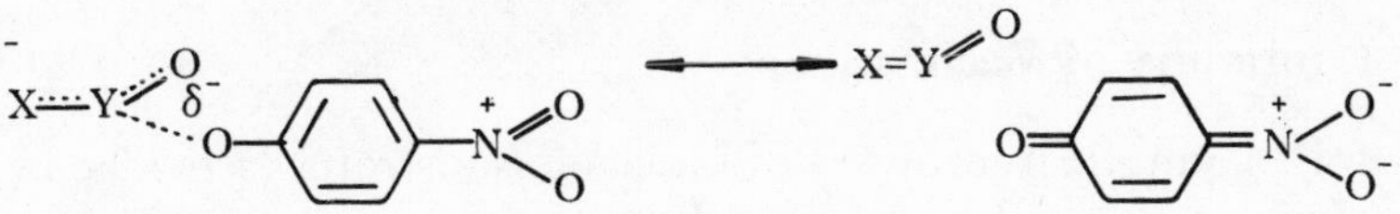

Scheme 2

Further evidence is that the entropies of activation are usually much more positive (often near zero) for E1cB reactions than for bimolecular models. As might be expected for a reaction in which the transition state is considerably fragmented, there need be little interaction with the nucleophile for productive reaction, as shown by the results of aminolysis studies on E1cB substrates. The nucleophilic reactions of amines with phosphoramidate [62], 4-nitrophenyl phosphate dianion [63], 2,4-dinitrophenyl phosphate dianion [64], and 4-nitrophenyl sulfate anion [32] exhibit low Brønsted values of 0.22, 0.13, 0, and 0.20, respectively, for amine attack ($\beta_{Nuc}$). The selectivity of 4-nitrophenyl sulfate towards thiophenols [65] can be calculated as approximately 0.05, although the ionization states of the reactants at 90°C in dimethylformamide are uncertain. There is essentially no dependence on amine structure or basicity in the aminolysis of 4-nitrophenyl methylaminosulfonate, even though at 1 $M$ concentration of amine all the product is in the form of sulfonamide [35]. Such observations indicate very little bond formation in the transition state for acyl transfer of such compounds compared with the high degrees of interaction for nonacidic centers: for the aminolyses of sulfonyl halides $\beta_{Nue}$ values range between 0.6 and 0.9 [66] and for P(V) phosphotriesters from 0.4 to 0.9 (for a discussion see reference [67]).

The transition state may indeed be coupled to some extent. The most obvious comparison is the very high $\beta_{LG}$ values and very low $\beta_{Nuc}$ values observed for E1cB reactions: for bimolecular reactions, the differences between $\beta_{Nuc}$ and $\beta_{LG}$ are less extreme. We can, however, consider the matter on a more detailed level. For substrates exhibiting relatively low (for E1cB reactions!) values of $\beta_{LG}$, the interaction with added nucleophiles (described by $\beta_{Nuc}$) is slight but observable. Thus for phosphate dianions and sulfate monoanions $\beta_{LG}$ is -1.2, with $\beta_{Nuc}$ = 0.13 for 4-nitrophenyl phosphate dianion and $\beta_{Nuc}$ = 0.20 for 4-nitrophenyl sulfate monoanion (Table 3). However, substrates showing very high values of $\beta_{LG}$ (for 4-nitrophenyl methylaminosulfonate $\beta_{LG}$ is -1.8) possess no kinetically observable dependence on amine basicity.

### 3.4 Continuity of Mechanism

In nonenzymatic hydrolyses of ionizable substrates (see Scheme 1), unimolecular elimination of the substrate anion ($k_1$) and bimolecular attack of water on this anion ($k_3$) should not be confused: they are distinct processes, albeit extremes. However, the continuum between uni- and bimolecularity exists as surely here as it does for $S_N1/S_N2$ reactions at $sp^3$ carbon. Nonetheless, for many of the reactions already described, the transition states have been shown unambiguously to have high degrees of E1 character (supported by large, negative $\beta_{LG}$ values and Hammett sigma-minus dependencies). In the extreme cases the E1cB route implies that water interaction in the transition state is limited to solvation of the intermediate—the transition state structure approximates that of the intermediate, say a planar metaphosphorimidate **VIII**. In the bimolecular attack of water on the conjugate base of the substrate, water is strongly bonded and the structure of the transition state reflects a tetrahedrally organized phosphorus center. However, the probes usually employed to implicate solvent participation in the transition state (activation parameter, isotope effects, solvent changes) are indefinitive in this area and, although they are useful to indicate *some* degree of interaction in the transition state, no quantitative description is yet possible. Even apart from the possibilities of partially rate-determining proton transfers (perhaps solvent mediated) in some cases, the distinction between differential solvation of transition and reactant states and specific nucleophilic solvent interaction as a fragmentation driving force may be semantic rather than real, in the twilight area of borderline mechanisms [31].

### 3.5 The Driving Forces for Elimination Reaction of Acyl Compounds

It has already been stated that the incursion of an E1cB route can lead to enormous rate increases (often of the order of $10^6$- to $10^8$-fold) over the corresponding bimolecular reaction. In this section the causes of these rate increases are discussed. It is noteworthy that the substrates concerned, for example, carbamates [22], are deactivated with respect to the electrophilicity of the acyl center (Scheme 3) and, if an E1cB mechanism were not available to circumvent the bimole-

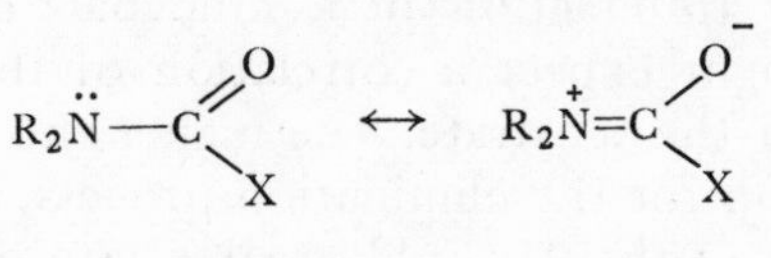

Scheme 3

cular route, such esters would not be nearly as labile in free solution as they are observed to be.

One may ask what the function of an enzyme is, if rate increases of the magnitudes mentioned can be obtained by a relatively straightforward mechanistic change. It must not be overlooked that a considerable part of the function of an enzyme is logistic—active sites serve as templates to bring two (or more) species together, apart from any other energetic considerations. In terms of biological acyl transfer, specificity is not recessive with respect to rate. Furthermore, the absolute, observed rates of many E1cB reactions, without some form of catalysis, are often low. One can quote the spontaneous hydrolysis of 4-nitrophenyl sulfate monoanion for which the rate constant at 35°C is about $2 \times 10^{-9}$ sec$^{-1}$ [32] while the acid-catalyzed rate constant for this ester at 25°C is about $2 \times 10^{-5}$ sec$^{-1}$ [33]. Intramolecular catalysis of sulfate hydrolysis is yet more efficient and carboxyl group catalysis increases the rate of hydrolysis of salicyl sulfate at pH 3 to 4 about 200-fold over that expected from hydronium ion catalysis [68]. Undoubtedly a large part of the function of aryl sulfatases is catalytic.

Below we consider the driving forces for the E1cB reaction of esters, and related species, in terms of the following:

1. Acidity of the substrate
2. The leaving group
3. Intermediate stability
4. The nucleophile.

*3.5.1 Acidity of the Substrate.* When there is no α-hydrogen atom available, as in fully *N*-alkylated carbamates or sulfamates, the E1cB route is obviously no longer possible. As the E1cB fragmentation step (rate-determining) involves an intramolecular nucleophilic attack on an acyl center (to give a π-bond), for example,

$$\overset{\curvearrowright}{RN}-SO_2-\overset{\curvearrowright}{X}$$

then, as far as "internal nucleophilicity" (here π-bond formation) can parallel basicity (interaction of a lone pair of electrons with a 1s orbital), one might expect a correlation of the rate of elimination with the $pK_a$ of the substrate. The higher the $pK_a$ of the ester, the better it is set up for the elimination process, all other things being equal. In this way the $10^7$-fold greater rate of cleavage of 4-nitrophenyl methylaminosulfonate anion X ($pK_{ester} \sim 9$) than of 4-nitrophenyl sulfate anion XI ($pK_{ester} \sim 2$) is readily explained [35]. The situation is not simple, however, since if a series of esters with

$$\text{Me}\overline{\text{N}}\cdot\text{SO}_2\text{OpNp} \qquad \overline{\text{O}}\cdot\text{SO}_2\text{OpNp}$$

$$\textbf{X} \qquad\qquad\qquad \textbf{XI}$$

varying leaving groups (and hence $pK_a$) is studied, the rate of elimination from the anion *decreases* with increased ester $pK_a$ (Fig. 1). Data available to date are limited to aryl methylaminosulfonates [35], aryl *N*-(4-nitrophenyl)-carbamates [22], and some phos-

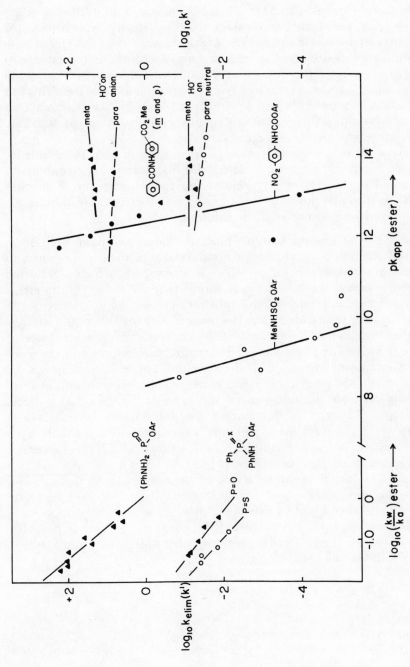

Figure 1.  Plots of log k′ (alkaline plateau rate constants) versus p$K_{substrate}$ for several ester series. References are given in the text.

213

phoramido esters [29-31]. Figure 1 shows a plot of log $K_{clim}$ (the rate constant for cleavage of the anion) versus $pK_a$ (of substrate) for these esters. The high dependence of the elimination rate for E1cB substrates contrasts sharply with the independence of substrate $pK_a$ of the rates of attack of hydroxide ion on the neutral and anionic forms of methyl 3- and 4-benzamidobenzoates. This last reaction has been shown not to be E1cB [69]. The influence of leaving-group ability enormously outweighs the effect of substrate $pK_a$ as can be seen from Figure 1.

As well as the complication of leaving-group activity, a balance must be struck between the need for high internal nucleophilicity, favored by high ester $pK_a$ values, and the requirement of an ester $pK_a$ sufficiently low for a kinetically sensible concentration of the conjugate base form to exist in solution.

*3.5.2 The Leaving Group.* Since the rate-determining step of an E1cB reaction is fission of the bond restraining the leaving group, an obvious prerequisite for an efficient cleavage is a highly activated leaving group. We have already shown that the leaving-group effect outweighs the acidity of the substrate in controlling reactivity. As the leaving ability decreases, we expect a change over to a bimolecular mechanism, in which nucleophilic interaction provides some of the driving force. This has been elegantly demonstrated in the work of Gerrard and Hamer [61] already cited. The leaving group may be activated, and the E1cB route catalyzed, by protonation, electrophilic catalysis (including metal ions), oxidation, or solvent effects (discussed later). As the leaving group is varied, the mechanism changes for poorer leaving groups—experimentally detectable by a variation in $\beta_{LG}$, for example, N-arylcarbamates [22], acetoacetates [24], and methylaminosulfonates [35]. This has been explained by a change to rate-determining attack of hydroxide ion on neutral ester for the acetoacetates and carbamates, but the sulfur work indicates that this is not always a safe conclusion in the absence of supporting evidence, because in this case [35] the driving force is probably provided for the poorer leaving groups by additional water interaction with the anionic ester XII.

$$
\begin{array}{c}
\delta- \\
\text{MeN} \cdots \overset{\delta-}{\underset{\delta-}{S}} \cdots \overset{O}{\underset{OAr}{}} \\
\delta-O \quad \\
H \quad H
\end{array}
$$

**XII**

*3.5.3* **Intermediate Stability Relative to Conjugate Base.** Most alkyl acetoacetates do not hydrolyze by the E1cB route [24], whereas ethyl *N*-phenylcarbamate has been shown to hydrolyze mainly by the EA path [21]. Increasing the thermodynamic stability of the intermediate, by going from a ketene to an isocyanate, makes the E1cB transition state energetically accessible, even for poorly activated ethyl esters. In agreement with this trend is the rapid solvolysis of alkyl carbonate monoesters [18], for which the required intermediate is carbon dioxide. This is a suitable point to clarify a potential source of confusion over the phrase "stability of the intermediate" by reference to Figure 2. As the intermediate is made *thermodynamically* more stable, its intrinsic energy decreases, with an accompanying decrease in $\Delta G^{\ddagger}$ (the distance between ground and transition states). However, it is not solely this process that governs the isolability or ease of trapping of the intermediate. The subsequent step may be rapid ($X_2 \rightarrow P_2$) or slow ($X_1 \rightarrow P_1$). Examples of the former case include the metaphosphate and isocyanate routes. The latter path may be formally exemplified by an olefin-forming E1cB reaction in aqueous solution, for which hydration of the

$\text{C}=\text{C}$ will not usually be fast in basic solution (eq. 12). The

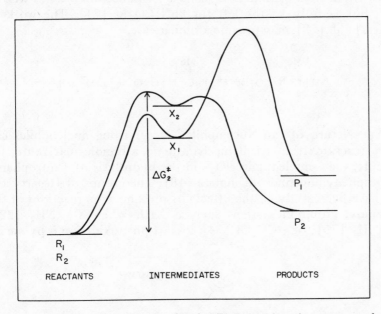

Figure 2. Formalized reaction profile for E1cB reactions in aqueous solution. $R_1$ and $R_2$ represent the conjugate bases of the substrate.

kinetic lability of the intermediate in solution must not be confused

$$\underset{\underset{}{}}{-}C-C-L \;\rightarrow\; \Large{>}\normalsize C=C\Large{<}\normalsize \;+\; L^- \;\xrightarrow[\;\times\;]{H_2O}\; \overset{OH}{\underset{H}{-C-C-}} \qquad (12)$$

with its thermodynamic stability—it is chiefly the former property that governs its reaction with added species or the possibility of its isolation.

In addition to the possibility of a relatively low-energy intermediate, a further propulsion for the E1 path lies in destabilization of the conjugate base of the substrate, for example, the presence of excess charge on phosphate dianions has been cited [63]. Entropically the E1 path is favored, especially for P-XYZ phosphorylating agents [70] (eq. 13), but resolution of the forming fragments sometimes

$$\overset{O}{\overset{\|}{O_2P}}-X-Y-Z \;\rightarrow\; [PO_3^-] \;+\; X{=}Y \;+\; Z^- \qquad (13)$$

counters this advantage. Steric crowding of the ester anion also favors E1 decomposition. For instance, equilibrium 14 lies well to right, but is fully to the left for $PhN \cdot CO_2Et$ [71]. The work of Remers et al. [72] may provide a similar case.

$$\overset{Me}{\underset{Et}{PhC-\overline{N} \cdot CO_2Et}} \;\rightleftharpoons\; \overset{Me}{\underset{Et}{PhC-N{=}C{=}O}} \;+\; EtO^- \qquad (14)$$

*3.5.4* **Nature of the Nucleophile.** Very strong nucleophiles can divert reaction from E1cB mechanism to a bimolecular route, for example, the α-nucleophile $HO_2^-$ in the hydrolysis of 4-nitrophenyl *N,N'*-diphenylphosphorodiamidate [48]. The groups of Hegarty and Fife have been studying the effects of proximity on reactions of the carbamoyl group in systems such as **XIII**. When X = $NH_2$ [22], $-CONH_2$ [75], or $-CH_2OH$[74], loss of phenoxide ion is by means

**XIII**

of a rapid E1cB reaction followed by intramolecular nucleophilic attack on the isocyanate with cyclization. Direct nucleophilic attack, without isocyanate formation, occurs for $X = CO_2^-$ [75] and $X = O^-$ [75, 76]. Thus the E1cB route seems energetically advantageous in all instances other than when a strong, charged, intramolecular nucleophile is present.

## 3.6  Catalysis of E1cB Routes of Ester Hydrolysis

This section briefly outlines the main ways in which the E1cB route can be catalyzed and thus is especially relevant to any discussion of the enzymatic reactions in which such substrates are involved. The most important source of catalysis appears to be protonation of the leaving group—especially when the proton can be supplied intramolecularly. However, metal ion, oxidative, and microsolvation (including micellar) effects have also been observed.

*3.6.1  Acid Catalysis.*  The promotion of E1 routes by zwitterion formation has been described for many reactions, including those of sulfamic acids [77, 78], the imidodisulfonate ion [79], aryl phosphate monoanions [26], phosphoroguanidates [42, 80], carbamylimidazoles [81, 82], phosphoramidates [83], S-phosphate monoanions [54], phosphonoformic acid [84], monosulfates [33, 85, 86], Bunte salts (RSSO$_3$M) [87], and N,N'-diarylsulfamides [88]. Zwitterions have been suggested in sulfene formation from alkanesulfonyl halides [53] and are presumably responsible for the rapid decomposition of carbamic acids in acidic solution [89,90].

Zwitterionization improves both the leaving group (p$K_a$) and the internal nucleophilicity of the $\alpha$-atom (eq. 15). This mechanism is

$$HO-SO_2OAr \rightleftharpoons {}^-O-SO_2-\overset{+}{\underset{H}{O}}Ar \tag{15}$$

only of value over the limited pH range in which the zwitterion concentration is kinetically sensible.

Phosphate monoesters commonly solvolyze more rapidly in the mono- rather than the dianionic form [26]. The rates of hydrolysis of the monoanions show a much lower dependence on leaving group ($\beta_{LG} = -0.3$ for O-phosphate monoesters [26,91] and 0.08 for S-phosphate monoesters [54]) than is observed for the corresponding dianions ($\beta_{LG} = -1.2$ for the O-phosphates [26,91]). This has been discussed, along with other other evidence, in terms of a preequilibrium proton transfer, which may become partially rate limiting for sufficiently active leaving groups [26], (see eq. 16). The decomposition of the zwitterion, designated $k_{elim}$, is analogous, in the absence of partial rate control by the proton transfer steps, to the unimolecular collapse of a phosphate dianion, albeit it now with

$$
\underset{\underset{OH}{|}}{RX \cdot \overset{O}{\overset{||}{P}} - O^-} \; \underset{}{\overset{K_{z\omega}}{\rightleftharpoons}} \; \underset{\underset{O}{\overset{|}{H}}}{RX - \overset{O}{\overset{||}{\underset{+}{P}}} - O^-} \; \overset{k_{elim}}{\longrightarrow} RXH \; + \; [PO_3^-]
$$

$$\Big\downarrow \text{fast} \Big| H_2O$$

$$RXH + H_2PO_4^-$$

(16)

a highly activated leaving group. As such, it would be expected to obey a similar linear free energy relationship (not identical with that of the dianions since the leaving group is formally neutral for the zwitterions but negatively charged for the dianions).

Other examples of intramolecular protonation of the departing function facilitating metaphosphate expulsion include the solvolysis of salicyl phosphate **XIV** dianion, in which proton transfer is little advanced ($\alpha = 0$), but P—O, cleavage is extensive ($\beta_{LG}$ is more comparable to the values of dianionic than monoanionic phosphate ester hydrolysis [92, 93]). It has been argued that a large rate difference between ortho and para carboxyphosphate esters reflects a large difference in the $pK_a$'s of the carboxyl groups [94] and on this basis the enhancements arising from general-acid catalysis for salicyl phosphate (**XIV**, $k_o/k_p = 200$, $\Delta pK = 3$), salicyl thiophosphate (**XV**, $k_o/k_p = 5$, $\Delta pK = 1.9$ [95]), and the lack of catalysis for phosphoenolpyruvate (**XVI** [96]), and 2-carboxyphenyl phosphoramidate (**XVII** [97]) have been explained [98].

XIV, Y = O                XVI
XV,  Y = S
XVII, Y = NH

Several other situations in which intramolecular protonation may augment leaving ability have been described in P(V) chemistry. Interesting intramolecular proton transfers have been suggested to account for the high tendencies of some acyl phosphates to undergo P—O fission as the monoanions, for example, acetyl phosphate (XVIII [43] and carbamyl phosphate XIX, [99]). The solvolysis of acetyl phosphate dianion occurs by P—O cleavage to a metaphosphate ion [43], but carbamyl phosphate dianions can react either by C—O fission (XX) to give isocyanate if unsubstituted [100], or by

XVIII, R = CH$_3$
XIX,   R = R'NH          XX

P—O scission analogous to acyl phosphate dianion hydrolysis if *N*-substituted [44].

Intramolecular general-acid catalysis of sulfate transfer has been observed, for example, salicyl sulfate (XXI [68]), 2-(4(5)-imidazolyl)phenyl sulfate (XXII [94, 101]), and 8-hydroxyquinoline sulfate (XXIII, [102, 103]. However, in none of these cases is the function of the proton facilitation clear—proton transfer to the leaving group may be aided or a zwitterionic transition state stabilized [10].

XXI          XXII          XXIII

*3.6.2 Electrophilic Catalysis.* Conceptually we might readily have treated protonic and metal ion catalyses together under the title of "electrophilic catalysis," since the suggestion is often made that metal ions act by virtue of their superacidity. However, it is convenient to allow proton catalysis as a separate class, the more so since metal ions, particularly transition-metal ions, have added properties useful in catalysis (variable oxidation state, potential action as templates, etc.).

Therefore, excluding the hydronium ion, the bulk of electrophilic catalysts is provided by metal ions. This is an area of great relevance to enzyme chemistry because the majority of phosphatases are metalloproteins and many kinases require metal cofactors. Magnesium ion is an essential cofactor in the enzymatic reactions of 3′-phosphoadenosine 5′-phosphosulfate [104, 105].

Several cases of metal-ion activated metaphosphate ion formation are available and all appear to involve at least bidentate chelation of the metal ion. As pointed out by Benkovic, rate increases (of the order of $10^5$-$10^6$) are directly proportional to the metal-ion-induced perturbation of the leaving group $pK_a$, and rates agree with those predicted on the basis of the perturbed $pK_a$ and the linear free energy relationship for phosphate monoester dianions [37].

The metal-ion-catalyzed hydrolysis of acyl phosphates has been explained [106] in terms of complexes such as XXIV, analogous to the explanation of proton-transfer reaction of the monoanion XVIII. There has been no demonstration of metal-ion-catalyzed metaphosphate elimination from polyphosphates, but in the cases studied the metal ion was shown to bind to a site other than the oxygen of the leaving group (XXV), a situation in which little catalysis of an E1 reaction would be expected.

XXIV                    XXV

However, $Mg^{2+}$ catalysis of phenyl phosphosulfate hydrolysis has been observed, and although details of the mechanism are not yet available (e.g., whether there is P—O or S—O fission), solvent polarity seems crucial to the observability of such an effect [107]. Metal-ion promotion has been observed in the hydrolysis of 8-hydroxyquinoline sulfate (XXVI) [102, 103].

Other forms of electrophilic catalysis have been reported. The hydrolysis of phosphoramidate is catalyzed by formaldehyde, hypo-

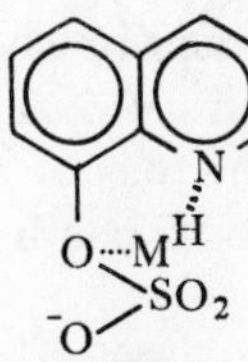

**XXVI**

chlorite, and nitrous acid [108] by electrophilic catalysis, wherein the catalyst, when covalently bound to the substrate, provides an improved leaving group (eq. 17). Hypobromous acid (but not

$$2^-O_3P \cdot NH_2 \longrightarrow 2^-O_3P \cdot NCl_2 \longrightarrow [PO_3^-] + NCl_2^- \qquad (17)$$

bromine) catalyzes the cleavage (**XXVII**) of bicarbonate ion [109].

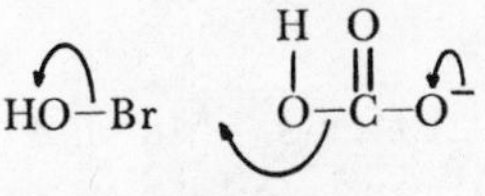

**XXVII**

*3.6.3* **Leaving-Group Oxidation.** Quinol phosphates, although resistant to hydrolysis, act as phosphorylating agents in the presence of oxidizing agents (e.g., $Br_2$); a metaphosphate pathway has been suggested [110] as in eq. 18. Analogous work has been successful on quinol sulfates [111].

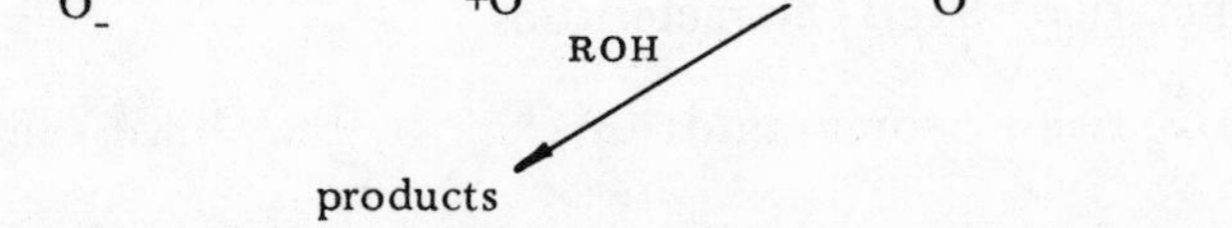

$$(18)$$

A similar suggestion [112] has been made for the oxidation of phosphite anion by iodine in alcohol to give quantitative yields of monoalkyl phosphates [113]. Although the molecularities of the above reactions are uncertain, it is interesting that the 6-chromanyl derivative of vitamin $K_1$, a hydroquinone phosphate, gives ATP in a purified bacterial oxidative-phosphorylation system when oxidized [114]. A phosphorylated form of NADH takes part in oxidative phosphorylation [115] by means of an intermediate (**XXVIII**) [116], in which the leaving group has been improved from a hydro-

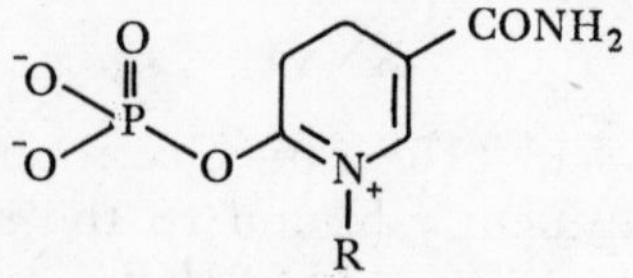

**XXVIII**

pyridine to a pyridinium species, promoting an E1cB reaction. Phosphorylation of NAD$^+$ would occur para to the amide with an oxidative activation step [117]. Phosphoryl derivatives of vitamin $K_2$ and ubiquinone have been suggested in phosphate transfer in the respiratory chain [118].

*3.6.4* **Microsolvation Effects.** Micellar catalysis of both phosphate ester anion [119] and sulfate ester anion [120] hydrolyses have been reported and cyclodextrin catalyzes aryl sulfate hydrolysis [121]. It is interesting that acid catalysis of sulfate hydrolysis becomes more pronounced in solvents of low polarity, especially ethers [122, 123]. Moist dioxane increases the rate of hydrolysis of alkyl sulfates approximately $10^7$-fold relative to water [123]. This is probably the result of specific solvation of the transition state, which for an E1 collapse of a zwitterion should resemble the product, sulfur trioxide (well known to give etherate [124, 125] complexes). In addition, low dielectric media tend to destabilize a zwitterionic ground state relative to the neutral, transition-state-resembling products of elimination. As the active sites of enzymes are likely to have hydrophobic sites, this contribution to activation should not be underestimated.

## 3.7 Summary of E1cB Characteristics

As an overview, compounds of class **II**, that have the following

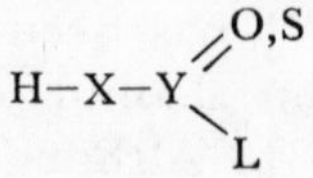

**II**

minimal requirements, are capable of using an E1cB mode of hydrolysis and benefitting from increased rates of reaction relative to the hypothetical bimolecular process.

1. The X—H bond is acidic: a compromise value of $pK_a$ is required to balance high nucleophilicity of $^-$X— with a reasonable concentration of substrate anion.

2. L is activated as a leaving group either by an inherently low $pK_a$ (for the conjugate acid LH) or by acid catalysis, electrophilic catalysis, and so forth.

3. The energy level of the required E1cB intermediate ($X=Y\overset{\diagup O}{}$) is accessible from the ground state (conjugate base), given fixed values of the parameters in 1 and 2.

Whether an E1cB reaction occurs, given these necessary, but not sufficient, prerequisites, depends on environmental influences, such as the power of any nucleophiles present (e.g., groups at enzyme active sites). However, the work of Hegarty and Fife and their co-workers has shown that even in the presence of powerful, *intramolecular* nucleophiles, such as the amino group, an E1cB path can still be preferable to nucleophilic reaction (cyclization), which argues strongly for its efficiency.

Model chemistry indicates that the above requirements are often fulfilled by the compounds of Table 1, and their analogs, in free solution and that their reactivities are dominated by such E1 transition states. The most characteristic properties of E1cB reactions are found in the transition states. Although structure and reactivity for E1cB substrate reactions vary enormously, the pattern established for the acyl transfer process is definite. Over a wide range of structural types, the transition state is highly dissociated, expulsion of the leaving group is *highly advanced*, and there is very little interaction with the ultimate acyl acceptor (Scheme 4) (whether water or an amine nucleophile).

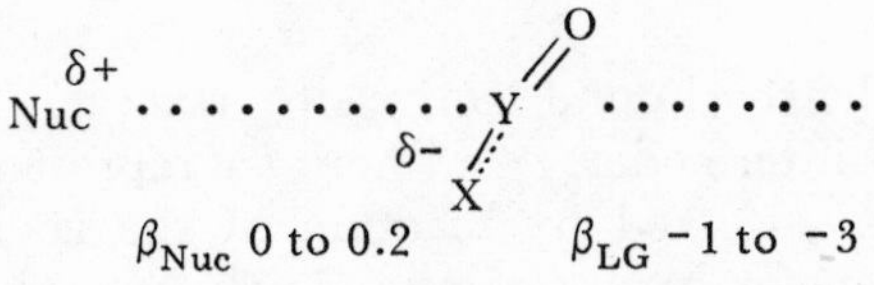

$$\underset{\beta_{Nuc}\ 0\ to\ 0.2}{\overset{\delta+}{Nuc} \cdots\cdots\cdots \overset{\overset{\textstyle O}{\diagdown\!\!\diagdown}}{\underset{\underset{X}{\delta-\ \diagup\!\!\diagup}}{Y}} \cdots\cdots\cdots \overset{\delta-}{L}\ \ \beta_{LG}\ ^-1\ to\ ^-3}$$

Scheme 4

This last point, weak interaction with a nucleophile (perhaps amounting to no more than solvation), is the basis of the low sensitivity of the reaction to nucleophiles (even intramolecular ones). This tendency implies that even at active sites of enzymes, unless binding specifically promotes a bimolecular process, one should expect much the same transition state characteristics as in free solution. The same could not have been said had the solution chemistry of the substrates indicated that passage across the reaction surface was dictated largely by the interaction with the nucleophile (as is the case for acetate esters, phosphinates, etc.). If the enzyme does force a substrate, which has a marked spontaneous tendency to form an E1 transition state, it must work against a rate factor often of the order of $10^6$ to $10^8$, and the advantage of the enzymatic process must lie very heavily in favor of some other aspect of the reaction.

## 4 SOME GENERAL FEATURES OF ENZYME INTERACTION WITH ANIONIC (E1cB) SUBSTRATES

In the following discussion the treatment is designed to show the nature of the processes possible for an ionized substrate in an enzyme active site. The enzyme is assumed to bind productively only to one ionic form of ionized substrate. No detailed kinetic analysis is presented, as observed parameters correspond to different rate constants for various enzymes in view of aggregation phenomena, multiple sites, and so on. However, an analysis of one particular case, the intermediacy of an acyl-enzyme, is given, since it is likely that acyl-enzymes are involved in the actions of many phosphatases and sulfatases.

Scheme 5 provides a generalized picture of *extremes* of interaction possible for an $\alpha$-acidic substrates (S) with an enzyme (E) that has an acceptor species (N) at the active site and also a general acid or metal ion ($BH^+$). Two limiting cases can be considered.

1. N is a noncovalently bound species (a complex between the enzyme and ultimate acyl acceptor is formed). In this case the $k_4$ step represents the dissociation equilibrium for the enzyme-products complexes.

2. N is a covalently bound active-site residue. This is so for acyl-enzyme (AcE) intermediacy and here $k_4$ represents deacylation steps. Hydrolysis is effected by reaction of the acyl-enzyme with water, and acyl transfer by interaction in the $k_4$ step with an acyl acceptor. Acyl transfer is also possible to a species other than N in the $k_3'$ (fast) step, if the E1cB route is followed.

In the first stage the EN species and S form a noncovalent com-

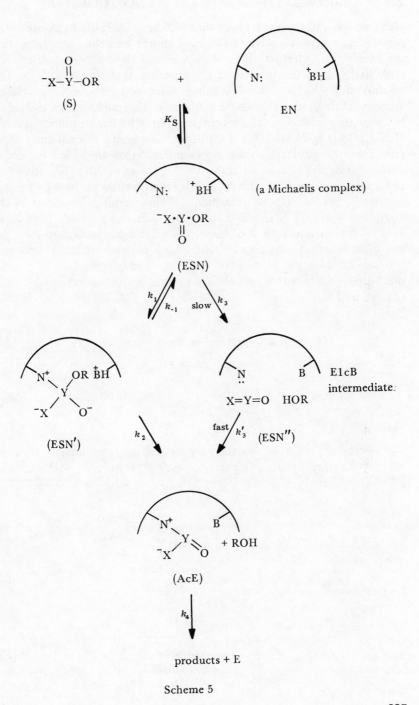

Scheme 5

225

plex, which can then react by two limiting routes; the $k_1$ route is the bimolecular attack of N on the acyl center and the transition state involves strong interaction with N, whereas the $k_3$ route is true ElcB with little or no interaction of the reacting acyl center with N. The second step $(k'_3)$ of the EA route is expected to be very fast by analogy with model studies of intramolecular nucleophilic attack on isocyanates. The specific interaction of the electrophilic catalyst $(BH^+)$ is not indicated for the bimolecular route since it may assist either at the acyl center by Y=O polarization or in leaving-group expulsion in the transition state. In the EA route the $BH^+$ species is only able to catalyze the leaving-group elimination in the $k_3$ step.

*Case 1.* A full kinetic treatment of this requires inclusion of the equilibria forming EN, and so on, and will not be given here as its exact form depends on the properties of the particular enzyme, for example, whether binding of N and S to E is sequential or random.

*Case 2.* This is analogous to the acyl-enzyme scheme fully described earlier for endopeptidases (e.g., $\alpha$-chymotrypsin) [126] (eq. 19 and 20).

$$E \; + \; S \xrightleftharpoons{K'_s} ES \xrightarrow{k_a} ES' \xrightarrow{k_d} E \; + \; P_2 \qquad (19)$$
$$+ \; P_1$$

$$k_{cat} \; = \; \frac{k_a k_d}{k_a + k_d} \qquad K_{m_{app}} \; = \; K'_s \; \frac{k_d}{k_a + k_d} \qquad (20)$$

However, $k_a$ must be replaced by a complex rate constant describing partitioning of the Michaelis complex between bimolecular and El routes, namely,

$$k'_a \; = \; k_3 \; + \; \left[ \frac{k_1 k_2}{k_{-1} + k_2} \right] \quad ; k_d = k_4 \qquad (21)$$

So that overall we may write:

$$k_{cat} \; = \; \frac{k_4 \left[ k_3 \; + \; \dfrac{k_1 k_2}{(k_{-1} + k_2)} \right]}{\left[ k_4 + k_3 \; + \; \dfrac{k_1 k_2}{(k_{-1} + k_3)} \right]} \qquad (22)$$

$$K_{m_{\text{app}}} = K_s \ \frac{k_4}{\left[k_3 + k_4 + \dfrac{k_1 k_2}{(k_{-1} + k_2)}\right]} \tag{23}$$

and

$$\frac{k_{\text{cat}}}{K_m} = \frac{\left[k_3 + \dfrac{k_1 k_2}{(k_{-1} + k_2)}\right]}{K_s} \tag{24}$$

For chymotrypsin, etc. $k_a$ and $K'_s$ have been obtained separately by study of the presteady state kinetics by means of rapid reaction techniques under substrate in excess conditions, using the usual Lineweaver-Burk separation. Thus for simple phosphoryl- and sulfuryl-enzyme intermediacy (discussed later), application of similar, presteady state analyses yields an experimental rate constant reflecting the properties of the acylation process [$k_3$ or $k_1 k_2/(k_{-1} + k_2)$]. Although the bimolecular route has been written as a stepwise process, in the limit when the intermediate shown (ESN′) is given zero lifetime ($k_2 \gg k_{-1}$), the acylation rate constant reduces to $k_1 + k_3$. It must be added that the best expression of E1cB characteristics, if any exist, will be found in a study of the presteady state. Equation 22 and 23 show that $k_{\text{cat}}$ and $(K_m)_{\text{app}}$, even if separated, are complex functions. Of the techniques that have been used in free solution to delineate E1 transition states, the most applicable is that of the linear free energy dependence for leaving-group change. Activation parameters and solvent isotope effects are ambiguous, even in the absence of enzyme. The observation of significant amounts of transacylation with no observable rate effects when the presteady state is studied may well be a useful approach, but has the added complication of acyl-acceptor binding. In addition, the rate of trans-phosphorylation must be comparable with the rate of acylation.

It is apparent that the forced proximity of N to the dissociating substrate in its transition state may emphasize the problem of continuity of mechanism between nucleophilic and solvation types of interaction of N with the reacting acyl center. However, this almost philosophical aspect of mechanistic distinction must await discussion until work specifically directed at active-site processes for E1cB transfer has appeared (some possible fruitful areas are mentioned below).

The above discussion concerned substrates for which the ionization process was diffusion controlled and the p$K$ (substrate) was in the normal aqueous solution range. However, the possibility that the enzyme lowers the p$K$ of a weakly acidic substrate has recently been suggested [10] for D-amino oxidase, which is inhibited by 4-nitrophenyl carbamate but not by ethyl carbamate [127]. The possibility of rate-determining proton transfer from carbon acids (e.g., acetoacyl coenzyme A derivatives) cannot be ignored in the enzymatic reactions of these important substrates.

## 5  ARYL SULFATASES

The values of $V_{max}$ (= $k_{cat}E_0$) for the hydrolyses of aryl sulfates catalyzed by several aryl sulfatases are independent of the nature of the substrate (as measured by p$K_{LG}$, for the phenol concerned), while $K_m$ values vary considerably, for example, for Type I [128] and II [129] aryl sulfatases of *Aspergillus oryzae* and the sulfatase A of ox liver [130]. The most likely explanation is that $V_{max}$ describes the slow desulfurylation of a sulfuryl-enzyme *($k_d < k_a$ in eq. 19 and 20)*, but the possibility of a rate-determining conformational change has not been disproved. On the basis of the intermediacy of an acyl-enzyme, one can readily explain the nonlinear dependence of log $V_{max}$ on p$K_{LG}$ in the reactions of aryl sulfates with Type II aryl sulfatase from *Alcaligenes metalcaligenes* [131] shown in Figure 3. For esters with activated leaving groups (low p$K_{LG}$), $V_{max}$ is approximately independent of p$K_{LG}$, but the value falls sharply as p$K_{LG}$ is increased, giving rise to a limiting value of $\beta_{LG} \approx -1.0$. One would expect by comparison with $a$-chymotrypsin [132] and papain [133], for which the acyl-enzyme path is well established, that, although $k_a > k_d$ for strongly acylating esters (low p$K_{LG}$), this will not be so for less reactive substrates, and $k_a$ (eq. 19) will gradually become rate determining, so that $V_{max}$ will reflect the acylation process. The approximate value of $\beta_{LG}$ of -1.0 for these sulfate esters in the enzymatic reaction is similar to the nonenzymatic hydrolysis of aryl sulfate anions ($\beta_{LG} = -1.2$), implying a transition state for the $k_a$ process with decided sulfur trioxide character. Values of log $(V_{max}/E \cdot K_m)(= \log k_a/K_s)$ have been correlated with the rates for nonenzymatic acid hydrolysis with the implication that the transition state has sulfur trioxide character [129]. However, the dangers of log-log correlations between enzymatic and nonenzymatic reactions are great, since binding effects are important for the former and steric effects on $k_a$ may be abnormal. However, until more is known about the effect of solvent and other factors on $\beta_{LG}$ values, direct comparison of Brønsted exponents from free solution with those from enzyme reactions is not safe.

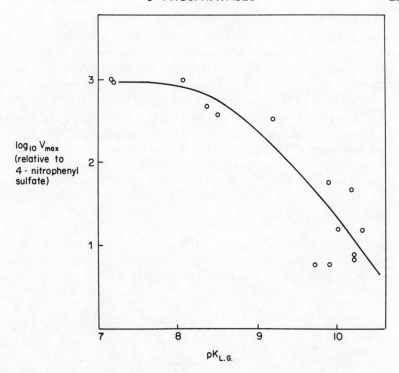

Figure 3. Brønsted-type plot of $\log_{10} V_{max}$ versus the $pK$ of the conjugate acid of the leaving group for hydrolysis of aryl sulfates by Type I aryl sulfatase from *Alcaligenes metalcaligenes*. Data were taken from reference 131.

# 6 PHOSPHATASES

Prostatic acid phosphatase hydrolyzes a series of phosphate mono-esters with constant $V_{max}$ but varying $K_m$ values [134], and it is tempting to use a treatment similar to that proposed for the sul-fatases—in this case rate-determining cleavage of a phosphoryl-enzyme. This is especially so since the enzyme-catalyzed hydrolysis of α-D-glucose-1-phosphate occurs with P—O cleavage, with no evidence of a pentacovalent intermediate and with a rate maximum at a pH where the monoanionic substrate predominates [135]. In addition, prostatic phosphatase catalyzes phosphate transfer not only to water but also to phenols, and so forth, at comparable rates [136, 137]. However, one cannot safely imply a metaphosphate-type transition state, as substrates are known for which $V_{max}$ varies considerably [138] and because, although almost all substrates are hydrolyzed at comparable rates at high pH by *E. coli* alkaline phosphatase, dephosphorylation here is not the rate-determining step and various conformational equilibria must be considered [139]. However, if mammalian acid phosphatase does indeed react by an acyl-enzyme route and $V_{max}$ reflects $k_a$ of eq. 19, the $\beta_{LG}$ value (obtained from a plot of log $V_{max}$ versus $pK_{LG}$ for the phenol) is $-0.2$ at pH $\geqslant 8$ [140]. At this pH the phosphates exist as dianions and a $\beta_{LG}$ of $-1.2$ would be expected (Table 3) for an E1 collapse to metaphosphate plus phenolate ions. The lower value of $\beta_{LG}$ is comparable with the value ($\beta_{LG} = -0.3$) found for the phosphate monoanion hydrolyses (Table 3), implying considerable proton transfer in the transition state.

# 7  DECARBOXYLASES

Formally the decarboxylation of the carboxylic acid can be written as in **XXIX**, which is analogous to the collapse of a monocarbonate to produce carbon dioxide. Although the intermediate involved here

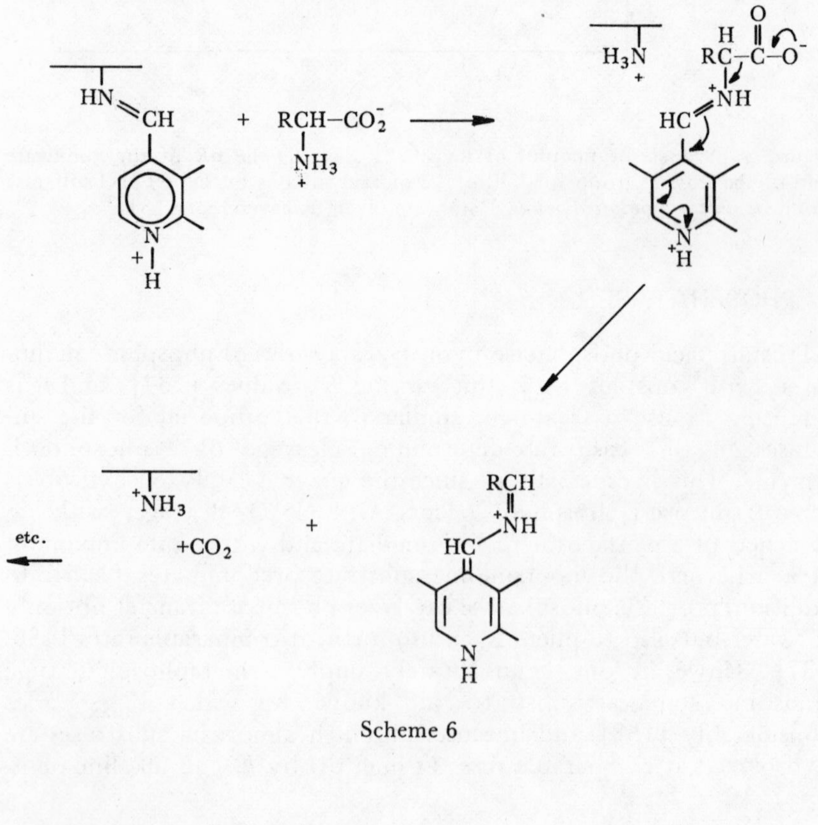

Scheme 6

$$R \cdot CH_2 - C \underset{O^-}{\overset{O}{\lessgtr}}$$

## XXIX

is of low energy and even alkyl carbonates can make use of an E1cB route, the leaving group in **XXIX** ($RCH_2-$) is too weak to allow E1 fragmentation. Transaldimination between an enzyme-pyridoxal 5′-phosphate complex and an amino acid substrate leads to a species that can decarboxylate readily [141, 142] in a reaction, which is, at least formally, E1cB, because the leaving group has been activated (see Scheme 6). Similarly, acetoacetate decarboxylase reacts by decarboxylation of a Schiff base [143].

## 8   REGULATION OF THE RATE OF ENZYME-CATALYZED ACYL TRANSFER

Some major points arise from the previous discussion of phosphate and sulfate transfer. The rule of these substrates over regulative metabolism is well known and the present work has shown how the aqueous solution transfers of these substrates possess a common and dominant feature. Most spontaneous phosphoryl transfers from substrates of type I occur by an elimination from an anionic species to produce a metaphosphate derivative (generally no more than incipient). The presence of strong nucleophiles, even in the same molecule, usually has little effect on the transferring transition state structure for such compounds. The distinctive feature of these anionic species is thus a seldom-circumvented, highly dissociated transition state, whose formation depends largely on the stability of the departing group. At an experimental level this manifests itself in a very high Brønsted exponent for leaving-group change. Table 3 indicates that this value is frequently near $-1.7$ for phosphate transfer,* between $-2$ and $-3$ for sulfonyl transfer, and approximate-

---

*For a monoacidic system reacting by Scheme 1, the apparent second order rate constant $k_{HO^-}$ is complex ($k_{HO} = k'K_a/K_w$ when [HO$^-$] is small). Consequently, the $\beta_{LG}$ value for $k_{HO^-}$ is ($\beta_{k'} + \beta_{K_a}$). For phosphate dianions $\beta_{k'} = -1.2$ (Table 3), but the value of $\beta_{K_a}$ is not known. However if we use the ionizations of aryl $N,N'$-diphenyl phosphoro- and phosphorothio-diamidates as models for which $\rho_{K_a}$ is $+1.27$ [29] and $+1.47$ [30], respectively, we can estimate $\beta_{K_a}$ using the $\rho$ for phenol ionization ($\beta_{K_a} = -\rho_{K_a}/2.2$) [144]. The value of $\beta_{K_a}$ thus obtained for these phosphorus esters is $+0.5$ to $+0.6$ giving $\beta_{HO^-} = -(+1.2+0.5)$, that is approximately $-1.7$ for phosphate monoesters. This is an approximation since we have considered only the monoacid level.

ly $-1.3$ for carboxyl transfer. It is these enormous changes in rate with leaving-group $pK_a$ (of conjugate acid) that give rise to the thesis of this section. We have seen that catalysts (e.g., general acids, metal ions) act by perturbing the $pK_{eff}$ of the departing species and produce their large rate increases because of the high Brønsted selectivities involved.

Consider a Brønsted equation (eq. 25) describing the dependency of the rate of elimination or of acyl transfer by an E1cB route on the $pK_a$ of the conjugate acid of the leaving group. $C$ is a constant. Decreasing the effective $pK_a$ of the leaving group by $n$ units leads

$$\log_{10}k = C + \beta pK_{LG} \qquad (25)$$

to a rate increase of $10^{n\beta}$ for acyl transfer.

The energy required for a group at the active site of an enzyme (e.g., a general-acid catalyst or a metal ion) to cause a decrease in the effective $pK$ of the leaving group of $n$ units is calculable from eq. 26.

$$\begin{aligned} \Delta G^{\ddagger} &= -RT \ln (\Delta K_a) \\ &= 2.303\, RT\, n \\ &\approx 1.36\, n \text{ kcal mole}^{-1} \text{ at } 25^{\circ}\text{C} \end{aligned} \qquad (26)$$

If we choose $\beta = -2.0$ in eq. 25 as a reasonable model (Table 3 supports this choice), we see that changing the effective $pK$ of the leaving group by 2 units leads to a $10^4$-fold increase in rate. The energy required for such a change is approximately 3 kcal mole$^{-1}$, well within the usual energy range of the hydrogen-bonding interaction. Thus hydrogen bonding, or indeed metal-ion complexation, at an active site could easily lead to very large rate changes in phosphate or sulfate transfer. One can look at model systems exhibiting intramolecular proton catalysis in this light (e.g., salicyl phosphate and sulfate).

Indeed, because of the critical dependence of the strength of the hydrogen bond on the distance between the electronegative atom centers (controlled by the A-H-O angle $\phi$ in XXX), a readily tunable

$$\Delta G^{\ddagger} \quad \text{maximum for } \Phi = 0^{\circ}$$
$$\Delta G^{\ddagger} \quad \text{minimum for } \Phi = 180^{\circ}$$

mechanism for varying the magnitude of the $pK$ perturbation, and hence the position of the transition state along the reaction coordinate, presents itself. There are many means of effecting such

**XXX**

angular control of leaving-group protonation, for example, competitive inhibition, metal-ion binding, allosteric effects. The overall result is to change the rate of the acyl-transfer step, with a direct effect on the overall rate if this happens to be the rate-determining step: if the acyl transfer is not the slowest step, the overall rate constant may still be significantly altered by a shift in the balance of rate constants for the sequence.

By means of this transition state "rheostat" effect the rate of a metabolically critical group transfer could be changed substantially by small energy changes, such as those expected from allosteric effectors, inhibitor binding, and so forth. A major difference between this and many other control schemes is that, rather than being an all or nothing effect, it is continuously variable. Hence a 50% decrease in the rate of a given step in a metabolic pathway is here caused by a 50% decrease in the rate of group transfer at the molecular level and not by complete inhibition of half of the enzyme molecules present.

## 9  CONCLUSION

It is hoped that this article has adequately demonstrated the similarity in mechanism that the many "high-energy" compounds of biochemistry possess, certainly as far as their free solution chemistry is concerned. It would be surprising if a feature as dominant as the similar, dissociated transition states that these substrates possess were not reflected to some considerable extent in enzyme mechanism, especially in view of the enormous rate, and consequently energetic, advantages offered by such E1cB routes. Indeed, this may be another example wherein evolution has come to a similar answer to a given problem, in this case group transfer, starting from different points— the similarity here not being in ground state structures, such as protein tertiary structures, but rather in the transition state properties of the substrates. Hopefully this article will stimulate direct attacks on this type of problem at the biochemical level, as the physical organic chemistry, although by no means complete, has already formed a solid foundation for experiment. The time has come for the work of model systems to be used pragmatically and to be applied directly to biochemical systems. While it is unlikely that physical organic chemistry will ever allow correct predictions of enzyme mechanism, it will certainly afford hypotheses that can be tested.

## ACKNOWLEDGEMENT

It is a pleasure to acknowledge Dr. A. Williams who first stimulated my interest in the area of elimination routes for ester hydrolysis and Drs. E. T. Kaiser and F. J. Kézdy for constructive critical comment and argument. I am indebted to Mrs. Hanna Posner for the formidable task of preparing this typescript.

I also wish to thank the University of Kent at Canterbury, the Northern Ireland Government, and the National Institutes of Arthritis, Metabolic, and Digestive Diseases for financial support.

## REFERENCES

1. D. E. Atkinson, *The Enzymes* (3rd ed.), 1, 474 (1970).

2. H. Holzer, *Adv. Enzymol.* (3rd ed.), 32, 297 (1969).

3. E. R. Stadtman, *Adv. Enzymol.* 28, 118 (1966).

4. S. A. Kuby and E. A. Noltman, *The Enzymes* (2nd ed.), 6, 515 (1962).

5. J. L. Dahl and L. E. Hokin, *Ann. Rev. Biochem.*, 43, 327 (1974).

6. P. F. Baker, *Metab. Pathways*, VI, 243 (1972).

7. F. Lipmann, *Adv. Enzymol.*, 6, 242 (1946).

8a. P. Reichard and G. Hanshoff, *Acta Chem. Scand.*, 10, 548 (1956).

8b. B. E. C. Banks in *Chemistry of the Amino Group*, S. Patai, Ed., Wiley, New York, 1968, p. 576.

9a. D. Gregory and F. Lipmann, *J. Biol. Chem.*, 229, 1081 (1957).

9b. E. Meezan and E. A. Davidson, *J. Biol. Chem.*, 242, 1685 (1967).

9c. A. S. Balasubramanian and B. K. Bachhawat, *Indian J. Exp. Biol.*, 1, 179 (1963).

10. A. Williams and K. T. Douglas, *Chem. Rev.*, 75, 627 (1975).

11. M. L. Bender, *Chem. Rev.*, 60, 53 (1960).

12. S. L. Johnson, *Adv. Phys. Org. Chem.*, 5, 237 (1967).

13. A. J. Kirby and S. G. Warren, *The Organic Chemistry of Phosphorus*, Elsevier, Amsterdam, 1967.

14. R. F. Hudson, *Structure and Mechanism in Organophosphorus Chemistry*, Academic Press, New York, 1965.

15. T. C. Bruice and S. J. Benkovic, *Bio-organic Mechanisms*, Vol. II, Benjamin, New York, 1966.

16. J. R. Cox, Jr. and O. B. Ramsay, *Chem. Rev.*, **64**, 317 (1964).
17. R. V. Vizgert, *Usp. Khim.*, **32**, 1 (1963).
18. N. F. Miller and L. O. Case, *J. Am. Chem. Soc.*, **57**, 810 (1935).
19. T. C. Bruice and S. J. Benkovic, *Bioorganic Mechanisms*, Vol. II, Benjamin, New York, 1966, p. 380.
20. L. W. Dittert, "The Kinetics and Mechanism of Base Catalyzed Hydrolysis of Organic Carbamates and Carbonates," Ph.D. Thesis, University of Wisconsin, 1961.
21. A. Williams, *J. Chem. Soc. Perkin II*, 808 (1972); 1244 (1973).
22. A. F. Hegarty and L. N. Frost, *J. Chem. Soc. Perkin II*, 1719 (1973).
23. M. L. Bender and R. B. Homer, *J. Org. Chem.*, **30**, 3975 (1965).
24. R. F. Pratt and T. C. Bruice, *J. Am. Chem. Soc.*, **92**, 5956 (1970).
25. C. A. Vernon, *Chem. Soc. (Lond.) Spec. Publ.*, **8**, 181 (1957).
26. A. J. Kirby and A. G. Varvoglis, *J. Am. Chem. Soc.*, **89**, 415 (1967).
27. P. S. Traylor and F. H. Westheimer, *J. Am. Chem. Soc.*, **87**, 553 (1965).
28. A. F. Gerrard and N. K. Hamer, *J. Chem. Soc. (B)*, **1968**, 539.
29. A. Williams and K. T. Douglas, *J. Chem. Soc. Perkin II*, 1454 (1972).
30. A. Williams and K. T. Douglas, *J. Chem. Soc. Perkin II*, 318 (1973).
31. A. Williams, K. T. Douglas, and J. S. Loran, *J. Chem. Soc. Perkin II*, 1010 (1975).
32. S. J. Benkovic and P. A. Benkovic, *J. Am. Chem. Soc.*, **88**, 5505 (1966).
33. E. J. Fendler and J. H. Fendler, *J. Org. Chem.*, **33**, 3852 (1968).
34. W. L. Matier, W. T. Comer, and D. Deitchman, *J. Med. Chem.*, **15**, 538 (1972).
35. A. Williams and K. T. Douglas, *J. Chem. Soc. Perkin II*, 1974, 1727.
36. A. Williams, K. T. Douglas and J. S. Loran, *Chem. Commun.*, **1974**, 689.
37. S. J. Benkovic and K. J. Schray, *Enzymes* (3rd ed), **8**, 201 (1973).
38. C. A. Bunton, *Accounts. Chem. Res.*, **3**, 257 (1960).
39. S. J. Benkovic in *Comprehensive Chemical Kinetics*, Vol. 10, C. H. Bamford and C. F. H. Tipper, Eds., Elsevier, Amsterdam, 1972.
40. A. Desjobert, *Colloq. Natl. Center Natl. Rech. Sci.*, **311**, 325 (1966).
41. The reason for this is discussed in reference 35.
42. P. Haake and G. W. Allen, *Proc. Natl. Acad. Sci. U.S.*, **68**, 2691 (1971).
43. G. DiSabato and W. P. Jencks, *J. Am. Chem. Soc.*, **83**, 4393, 4400 (1961).
44. C. M. Allen, Jr. and J. Jamieson, *J. Am. Chem. Soc.*, **93**, 1434 (1971).
45. S. J. Benkovic and R. C. Hevey, *J. Am. Chem. Soc.*, **96**, 4971 (1970).
46. W. Tagaki, T. Eiki and I. Tanaka, *Bull. Chem. Soc. Japan*, **44**, 1139 (1971).
47. K. T. Douglas, Ph.D. Thesis, University of Kent at Canterbury, 1973.
48. D. B. Coult and M. Green, *J. Chem. Soc.*, **1964**, 5478.
49. A. J. Kirby and G. J. Lloyd, *Chem. Commun.*, **1971**, 1538.
50. J. Casanova, Jr. and D. A. Rutolo, Jr., *J. Am. Chem. Soc.*, **91**, 2347 (1969).
51. E. T. Kaiser, P. Müller, O. R. Zaborsky, and D. F. Mayers, *J. Am. Chem. Soc.* **91**, 6732 (1969).
52. J. F. King, *Acc. Chem. Res.*, **8**, 10 (1975).

53. J. F. King, E. A. Luinstra, and D. R. K. Harding, *Chem. Commun.*, 1972, 1313.

54. S. Milstein and T. H. Fife, *J. Am. Chem. Soc.*, **89**, 5820 (1967).

55. G. Sartore, M. Bergon, and J-P. Calmon, *Tetrahedron Lett.*, 3133 (1974).

56. I. Dostrovsky and M. Halmann, *J. Chem. Soc.*, **1953**, 508.

57. P. A. T. Swoboda, *Chem. Soc. (Lond) Spec. Publ.*, **8**, 41 1957.

58. J. D. Chanley and E. Feageson, *J. Am. Chem. Soc.*, **85**, 1181 (1963).

59. W. P. Jencks, *Chem. Rev.*, **72**, 7053 (1972); *J. Am. Chem. Soc.*, **94**, 4731 (1972).

60. L. L. Schalager and F. A. Long, *Adv. Phys. Org. Chem.*, **1**, 1 (1963).

61. A. F. Gerrard and N. K. Hamer, *J. Chem. Soc. (B)*, **1967**, 1122.

62. W. P. Jencks and M. Gilchrist, *J. Am. Chem. Soc.*, **87**, 3199 (1965).

63. A. J. Kirby and W. P. Jencks, *J. Am. Chem. Soc.*, **87**, 3209 (1965).

64. A. J. Kirby and A. G. Varvoglis, *J. Chem. Soc. (B)*, **1968**, 135.

65. T. Kurusu, W. Tagaki, and S. Oae, *Bull. Chem. Soc. Japan*, **43**, 1553 (1970).

66a. O. Rogne, *J. Chem. Soc. (B)*, **1969**, 663; **1970**, 1056; **1971**, 1334.

66b. O. Rogne, *J. Chem. Soc. Perkin II*, **1972**, 472.

67. S. A. Khan and A. J. Kirby, *J. Chem. Soc. (B)*, **1970**, 1172.

68. S. J. Benkovic, *J. Am. Chem. Soc.*, **88**, 5511 (1966).

69. A. Williams and K. T. Douglas, *J. Chem. Soc. Perkin II*, **1972**, 2112.

70. V. M. Clark and D. W. Hutchinson, *Prog. Org. Chem.*, **7**, 75 (1968).

71. D. J. Woodcock, *Chem. Commun.*, **1968**, 267.

72. W. A. Remers, R. H. Roth, and M. J. Weiss, *J. Org. Chem.*, **30**, 2910 (1965).

73. A. F. Hegarty, L. N. Frost, and J. H. Coy, *J. Org. Chem.*, **39**, 1089 (1974).

74. J. E. C. Hutchins and T. H. Fife, *J. Am. Chem. Soc.*, **95**, 3786 (1973).

75. A. F. Hegarty, L. N. Frost and D. Cremin, *J. Chem. Soc. Perkin II*, **1974**, 1249.

76. J. E. C. Hutchins and T. H. Fife, *J. Am. Chem. Soc.*, **95**, 2282 (1973).

77. W. J. Spillane, *Int. J. Sulfur Chem.*, **8**, 469 (1973).

78. W. J. Spillane and F. L. Scott, *Mech. React. Sulfur Compd.*, **5**, 59 (1970).

79. G. J. Doyle and N. Davidson, *J. Am. Chem. Soc.*, **71**, 3491 (1949).

80. V. M. Clark and S. G. Warren, *Nature*, **199**, 657 (1963).

81. A. F. Hegarty, C. N. Hegarty, and F. L. Scott, *J. Chem. Soc. Perkin II*, **1974**, 1258.

82. A. Williams and W. P. Jencks, *J. Chem. Soc. Perkin II*, 1753, 1760 (1974).

83. T. C. Bruice and S. J. Benkovic, *Bio-organic Mechanisms*, Vol. 1, Benjamin, New York, 1966, p. 73.

84. S. G. Warren and M. R. Williams, *J. Chem. Soc. (B)*, **1971**, 618.

85. P. D. Batts, *J. Chem. Soc. (B)*, **1966**, 547.

86. J. L. Kice and J. M. Anderson, *J. Am. Chem. Soc.*, **88**, 5242 (1966).

87. J. L. Kice, J. M. Anderson, and N. E. Pawlowski, *J. Am. Chem. Soc.*, **88**, 5245 (1966).

88. W. J. Spillane, J. A. Barry and F. L. Scott, *J. Chem. Soc. Perkin II*, **1973**, 481.

89.  S. L. Johnson and D. L. Morrison, *J. Am. Chem. Soc.*, **94**, 1323 (1972).

90.  M. Caplow, *J. Am. Chem. Soc.*, **90**, 6795 (1968).

91.  C. A. Bunton, E. J. Fendler, E. Humeres, and K.-U. Yeng, *J. Org. Chem.*, **32**, 2806 (1967).

92.  M. L. Bender and J. M. Lawlor, *J. Am. Chem. Soc.*, **85**, 3010 (1963).

93.  R. H. Bromilow and A. J. Kirby, *J. Chem. Soc. (B)*, 1972, 149.

94.  S. J. Benkovic and L. K. Dunikoski, *Biochemistry*, **9**, 1390 (1970).

95.  T. H. Fife and S. Milstein, *J. Org. Chem.*, **34**, 4007 (1969).

96.  S. J. Benkovic and K. J. Schray, *Biochemistry*, **7**, 4090 (1968).

97.  S. J. Benkovic and P. A. Benkovic, *J. Am. Chem. Soc.*, **89**, 4714 (1967).

98.  S. J. Benkovic and K. J. Schray, *Enzymes* (3rd ed.) **8**, 201 (1973).

99.  C. M. Allen, Jr., E. Richardson and M. E. Jones in *Current Aspects of Bioenergetics*, N. O. Kaplan and E. P. Kennedy, Eds., Academic Press, New York, 1966, p. 401.

100.  C. M. Allen, Jr., and M. E. Jones, *Biochemistry*, **3**, 1238 (1964).

101.  B. E. Fleischfresser and I. Lauder, *Aust. J. Chem.*, **15**, 243, 251 (1962).

102.  R. W. Hay and J. A. G. Edmonds, *Chem. Commun.*, 1967, 969.

103.  R. W. Hay, C. R. Clark, and J. A. G. Edmonds, *J. Chem. Soc. Dalton*, 1974, 9.

104.  F. Lipmann, *Science*, **130**, 1319 (1959).

105.  A. B. Roy, *Adv. Enzymol.*, **22**, 204 (1960).

106.  J. P. Klinman and D. Samuel, *Biochemistry*, **10**, 2126 (1971).

107.  W. Tagaki, Y. Asai, and T. Eiki, *J. Am. Chem. Soc.*, **95**, 3037 (1973).

108.  W. P. Jencks and M. Gilchrist, *J. Am. Chem. Soc.*, **86**, 1410 (1964).

109.  M. Caplow, *J. Am. Chem. Soc.*, **93**, 230 (1971).

110.  V. M. Clark, D. W. Hutchison, A. J. Kirby, and A. Todd, *J. Chem. Soc.*, 1961, 715.

111.  S. W. Weidman, D. F. Mayers, O. R. Zaborsky, and E. T. Kaiser, *J. Am. Chem. Soc.*, **89**, 4555 (1967).

112.  A. J. Kirby and S. G. Warren, *The Organic Chemistry of Phosphorus*, Elsevier, Amsterdam, 1967, p. 369.

113.  A. J. Kirby, *Chem. Ind. (London)*, 1963, 1877.

114.  A. Asano, A. F. Brodie, A. Wagner, P. E. Wittreich, and K. A. Folkers, *Fed. Proc.*, **21**, 54 (1962).

115.  D. E. Griffiths, *Fed. Proc.*, **22**, 1064 (1963).

116.  J. A. Barltrop, P. W. Grubb, and B. Hesp, *Nature*, **199**, 759 (1963).

117.  V. M. Clark, D. W. Hutchinson, and D. E. Wilson, *Angew. Chem.*, **77**, 259 (1965).

118a.  E. Lederer and M. Vilkas, *Vitam. Horm.*, **24**, 409 (1966).

118b.  V. M. Clark and A. Todd, *Quinones and Electron Transport*, M. O'Connor and G. E. W. Wolstenholme, Eds., Churchill, London, 1961, p. 190.

118c.  E. Lederer and M. Vilkas, *Exper.*, **18**, 546 (1962).

118d.  V. M. Clark, *Mechanismen Enzymatischer Reactionen*, Springer-Verlag, 1964, p. 276.

119.  C. A. Bunton, E. J. Fendler, L. Sepulveda, and K-U. Yang, *J. Am. Chem. Soc.*, **90**, 5512 (1968).

120. E. J. Fendler, R. R. Liechti, and J. H. Fendler, *J. Org. Chem.*, **35**, 1658 (1970).
121. W. I. Congdon and M. L. Bender, *Bioorganic Chem.*, **1**, 424 (1971).
122. S. Burstein and S. Lieberman, *J. Am. Chem. Soc.*, **80**, 5235 (1958).
123. B. D. Batts, *J. Chem. Soc. (B)*, **1966**, 551.
124. C. M. Suter, P. B. Evans, and J. M. Kiefer, *J. Am. Chem. Soc.*, **60**, 538 (1938).
125. C. M. Suter, *The Organic Chemistry of Sulfur*, Wiley, New York,1944, pp. 1-94.
126. M. L. Bender, F. J. Kézdy, and F. C. Wedler, *J. Chem. Ed.*, **44**, 84 (1966).
127. R. A. Abeles, quoted in A. Williams and K. T. Dougles, *Chem. Rev.*, (1975).
128. D. Robinson, J. N. Smith, B. Spencer, and R. T. Williams, *Biochem. J.*, **51**, 202 (1952).
129. S. J. Benkovic, E. V. Vergera, and R. C. Hevey, *J. Biol. Chem.*, **246**, 4926 (1971).
130. R. G. Nicholls and A. B. Roy, *The Enzymes* (3rd ed.), 5, 34 (1971).
131. K. S. Dodgson, B. Spencer, and K. Williams, *Biochem. J.*, **64**, 216 (1956).
132. M. L. Bender, G. R. Schonbaum, and B. Zerner, *J. Am. Chem. Soc.*, **84**, 2540 (1962).
133. J. F. Kirsch and M. Igelström, *Biochemistry*, 5, 783 (1966).
134. G. S. Kilsmeier and B. Axelrod, *J. Biol. Chem.*, **227**, 879 (1957).
135. M. Cohn, *J. Biol. Chem.*, **180**, 771 (1949).
136. J. Appleyard, *Biochem. J.*, **42**, 596 (1948).
137. G. Schmidt, *The Enzymes* (2nd ed.), 5, 40 (1961).
138. G. Bartsch, S. J. Thannhauser and G. Schmidt, *Fed. Proc.*, **19**, 332 (1960).
139. I. B. Wilson, J. Dayan, and K. Cyr, *J. Biol. Chem.*, **239**, 4182 (1964).
140a. Using results of G. E. Delory and E. J. King, *Biochem. J.*, **37**, 547 (1943).
140b. Using results of P. G. Walker and E. J. King, *Biochem. J.*, **47**, 93 (1950).
141. D. Metzler, M. Ikawa, and E. E. Snell, *J. Am. Chem. Soc.*, **76**, 648 (1954).
142. F. H. Westheimer, quoted in S. Mandeles, R. Koppelman, and M. E. Hanke, *J. Biol. Chem.*, **209**, 327 (1954).
143. W. Tagaki and F. H. Westheimer, *Biochemistry*, 7, 901 (1968).
144. G. B. Barlin and D. D. Perrin, *Quart. Rev.*, **20**, 820 (1966).
145. D. E. Wiggins, private communication quoted in K. T. Douglas, Ph.D. Thesis, University of Kent at Canterbury, 1973.
146. T. R. Fukuto and R. L. Metcalf, *J. Agr. Food Chem.*, **4**, 930 (1956).
147. I. Christianson, *Acta Chem. Scand.*, **18**, 904 (1964).

# HYDROLYSIS OF
# CYCLIC ESTERS

## E. T. KAISER AND F. J. KÉZDY

*Departments of Biochemistry and Chemistry*
*University of Chicago*
*Chicago, Illinois*

## 1  INTRODUCTION

The formation of a covalent bridge between the acyl and alkyl moieties of an organic carboxylic ester results in a profound alteration of the stereochemistry of the ester function. Unless the bridge contains more than 8 to 10 carbon atoms, the alkyl and acyl groups are forced into a *cis* configuration around the acyl carbon-alkyl oxygen linkage, whereas the open esters exist exclusively in the *trans* configuration. Ring formation also introduces marked deviations from the coplanarity normally found for open esters. In addition to these stereochemical changes, the cyclic structure also imposes restrictions on the relative movements of the alkyl and acyl positions of the molecule, resulting in entropic changes; and if the ring is strained, large enthalpic changes are also manifest between the cyclic ester and its open-chain analog.

In the light of these structural and thermodynamic differences, it is not surprising that the rate of the hydrolytic ring opening can be different by several orders of magnitude from that of the hydrolysis of the corresponding open-chain esters. The rate of hydrolytic ring opening is governed by the free energy difference ($\Delta F_c{}^{\ddagger}$) between the ground state and the transition state, which is also cyclic. Therefore the comparison of the hydrolytic rates of the cyclic ester and its open-chain analog ($\Delta F_c{}^{\ddagger} - \Delta F_o{}^{\ddagger}$) reflects not only the free energy changes of the ground states upon cyclization ($\Delta F$), but also those between the transition states ($\Delta F'$), that is, $\Delta F_c{}^{\ddagger} - \Delta F_o{}^{\ddagger} =$

$$\Delta F - \Delta F' = RT \ln \frac{k_c}{k_o}.$$ Thus the rate comparisons between cyclic and

open-chain esters do not yield results directly interpretable in terms of the structure of ground state, and one cannot separate the discussion of the stereochemistry and the energetics of cyclic esters from those of the cyclic transition states. The interpretation of the kinetic results is further complicated because the conformation and configuration of the cyclic esters may also favor hydrolytic mechanisms that are normally not observed in open-chain esters.

This latter possibility greatly stimulated research in the hydrolytic mechanisms of cyclic esters, especially after the discovery of a host of biologically important cyclic esters and amides, $2',3'$-cyclic nucleotides, cyclic AMP, pyroglutamic acid, penicillin, and homocysteine thiolactone, to name only a few. Also, the possibility of rate acceleration through strain made cyclic esters the favorite models for enzyme mechanistic studies.

## 2 PRELIMINARY CONSIDERATIONS

The large majority of ester hydrolyses occur by the addition-elimination pathway, rather than by elimination-addition or by $S_N2$ single-step bimolecular collision. Cyclic esters in general follow the same pathway, with the formation of a tetracoordinated intermediate in the hydrolysis of carboxylic esters and a pentacoordinated one in the case of cyclic phosphates and possibly sulfates. For carboxylic esters the addition-elimination pathway presents an obvious advantage with respect to the other two mechanisms in that the resonance stabilization of the ester function is disrupted in the addition step without breaking the bond between the acyl group and the leaving group. Thus the two energetically unfavorable processes occur in two distinct steps, thereby lowering the activation energy of the overall process. At the same time the formation of the addition intermediate results in the transformation of the coplanar ester function into a tetrahedral dihydroxy ether. While noncyclic ester structures readily allow such a movement of the alkyl group with respect to the acyl moiety—with the exception perhaps where extreme steric hindrance exists in the two groups— the ring structure imposes severe constraints on the stereochemistry of the groups attached to the cyclic carbonyl carbon atom. Thus *a priori* one would expect that steric hindrances and the weakening of bonds by steric strain would play a prominent role in the hydrolytic opening of the ring.

Since the addition of the nucleophile results only in the rearrangement of the stereochemistry of the ring but not in its opening, the free energy changes related to the appearance of free rotation in the noncyclic structure manifest themselves only in the elimination step. The overall rate $(V)$ of nucleophilic ring opening of the ester (E) by the nucleophile (N), as represented by the scheme of eq. 1 is given by eq. 2. Thus the accelerations due to chain randomization are

$$E + N \underset{k_{-1}}{\overset{k_1}{\rightleftharpoons}} I \overset{k_2}{\longrightarrow} P \tag{1}$$

$$V = \frac{k_1 k_2}{k_{-1} + k_2} [E] [N] \tag{2}$$

apparent only if $k_{-1} > k_2$, that is, in the case of poor leaving groups in the ester function and when the nucleophiles themselves are good leaving groups.

It has been estimated that the freezing of the free rotation around a carbon-carbon bond in the transition state requires an activation entropy of the order of 5 E.U. An accelerating effect of a similar order of magnitude per C—C bond (~200-fold) is expected then for the opening of a rigid alicyclic ring. Since increasing the number of methylene groups in the ring decreases its rigidity at the same time, one should expect that rate acceleration due to this "entropy effect" would level off at larger ring sizes, or even pass through a maximum in the vicinity of six- or seven-member rings.

Since the entropy effect results in the selective acceleration of the ring-opening step, it favors elimination reactions over addition reactions. Thus, in appropriate cases, one should observe a change in mechanism from addition-elimination to elimination-addition, especially in the case of small, rigid alicyclic rings with poor leaving groups.

Finally, cyclic esters differ markedly from their open-chain analogs in that the in the hydrolysis product the alcohol is confined to the immediate neighborhood of the acid function. Because of this "local concentration" effect the reverse reaction of ester formation can be greatly facilitated, and the thermodynamic equilibrium between the ester and the acid might even lie completely on the ester side, unless the ionization of the acid displaces the equilibrium toward the acid side.

In summary, the study of the hydrolysis of cyclic esters in comparison with their open-chain analogs should not only lead to the elucidation of the structural features of small rings but also to a better understanding of the mechanism of ester hydrolysis. In the following pages we hope to show that indeed some progress has been made in this direction.

Because of the complex relationship between their structure and their reactivity, however, cyclic esters are slow to yield their secrets. Thus the present article is more of a progress report than a definite treatise of their hydrolytic mechanisms. In the following pages we discuss the present knowledge concerning the hydrolysis of cyclic phosphates, sulfates, and lactones, the best known examples of the cyclic ester family. At the end, the relevancy of these results to biochemical and enzymatic mechanisms is briefly evaluated.

## 3  CYCLIC ESTERS OF PHOSPHORIC ACIDS

Since excellent reviews on the reactivity of cyclic phosphates have been published elsewhere [1,2], this topic is treated here only briefly. Most attention in the hydrolysis of cyclic phosphates has been devoted to the reactions of the five-membered ring species. This is due to the intermediacy of such esters in the alkaline and enzymatic hydrolysis of the ribonucleic acids and to their exceptional lability to hydrolysis [1-3]. Specifically, the simple five-membered cyclic phosphate diester ethylene phosphate hydrolyzes in alkali at a rate approximately $10^7$ times as great as the hydrolysis rate of corresponding acyclic phosphate diesters like dimethyl phosphate under comparable conditions. Similarly, $2',3'$-cyclic nucleotides are hydrolytically labile componds [4]. When the observations that ethylene phosphate hydrolyzes under alkaline conditions with exclusive P—O bond cleavage while dimethyl phosphate undergoes mainly C—O cleavage [5] is taken into account, the rate enhancement for attack at the phosphorus atom of the five-membered cyclic ester by hydroxide ion can be estimated to be greater than $10^8$. In contrast, six-membered cyclic phosphate diesters, such as trimethylene phosphate [6] and $3',5'$-cyclic AMP [7] hydrolyze at rates comparable to or only slightly faster than those of the simple acyclic phosphate diesters. The effects of strain have been implicated in the rapid hydrolysis of the five-membered cyclic phosphate esters, and evidence for such strain has come from measurements on the enthalpies of hydrolysis of labile triesters of $2',3'$-cyclic nucleotides such as methyl ethylene phosphate, [9], and recently from measurements of the heat of hydrolysis of the diester sodium ethylene phosphate catalyzed by a new phosphodiesterase prepared from *Enterobacter aerogenes* [10]. The enthalpies of hydrolysis of sodium diethyl phosphate and of sodium trimethylene phosphate have been found to be smaller than that of sodium ethylene phosphate by about 4 kcal mole$^{-1}$. In line with the thermochemical measurements on the hydrolysis of the five-membered cyclic phosphate esters, the x-ray crystallographic structure determinations performed on a number of the five-membered cyclic species [11, 12] suggest that there is considerable angle strain in the ring. Specifically, very small internal O—P—O bond angles have been found for the five-membered rings, and it has been proposed that angle strain represents a significant part of the overall strain energy of these systems [13].

As in base, the acidic hydrolysis of the five-membered cyclic esters of phosphoric acid proceeds millions of times faster than that of the related acyclic species [3]. An important observation made in acidic solutions is that the hydrolysis of hydrogen ethylene phosphate is accompanied by rapid oxygen exchange from the aqueous solvent into the unreacted ester [5]. Along similar lines, it has been observed that hydrolytic cleavage of the methoxyl group external to the ring in methyl ethylene phosphate, a triester that is hydrolyzed in acid approximately $10^6$ times as fast trimethyl phosphate, occurs at a rate that is competitive with cleavage of the ring [14, 15]. While the rapid opening of the ring in the five-membered cyclic phosphates is consistent with the observation of ring strain, an important question is raised by the oxygen-exchange experiment on hydrogen ethylene phosphate and the observation of methoxyl cleavage in the hydrolysis of methyl ethylene phosphate: how can strain accelerate exchange or hydrolysis in these systems without the opening of the ring? On the basis of a detailed analysis of a variety of reactions of cyclic phosphorus-containing esters, Westheimer and his co-workers have presented a forceful case for the hypothesis that hydrolysis external to the five-membered ring or the exchange of the phosphate oxygen atoms proceeds with "pseudorotation" between trigonal-bipyramidal intermediates [3, 15-17]. Because the arguments in favor of the postulation of trigonal-bipyramidal intermediates in phosphate ester hydrolysis have been presented with great elegance elsewhere, this topic is not discussed here. However the nature of these intermediates is relevant to cyclic sulfate and sulfonate hydrolysis reactions and in the part of this chapter dealing with those compounds the question of the existence of pentacoordinate intermediates and their structure is reopened.

Most of the studies performed on five-membered cyclic phosphate ester hydrolyses have been concerned with the reactions of aliphatic species. However, it has been demonstrated that the aromatic five-membered cyclic phosphate diester catechol cyclic phosphate undergoes alkaline hydrolysis almost $10^7$ times as fast in alkali as does its open-chain analog diphenyl phosphate [18]. An x-ray structure determination [19] of the bond angles and bond lengths in catechol cyclic phosphate has shown that the endocyclic O—P—O bond angle in this compound is $98.4 \pm 0.2°$, a value close to those observed for labile five-membered cyclic phosphate triesters [11, 12, 20, 21]. No thermochemical data have been reported for the hydrolysis of catechol cyclic phosphate nor has the rate of hydrolysis in acidic medium been compared to the rate of oxygen exchange from the solvent into the phosphate group of the unreacted cyclic ester. Nevertheless, the observation of a very small internal O—P—O bond angle in the five-membered ring of the aromatic cyclic phosphate ester suggests that ring strain probably plays a significant role in the enhancement of hydrolytic reactivity, as appears to be the case with the corresponding aliphatic systems.

While the situation regarding our understanding of the hydrolytic stability of five-membered cyclic phosphate esters seems reasonably well in hand, some rather perplexing observations made on six-membered cyclic phosphate esters still need to be explained. Although the six-membered cyclic phosphate diesters trimethylene phosphate and 3',5'-cyclic AMP are not very labile to hydrolysis and behave similarly in this regard, there is an enormous difference between these compounds when their enthalpies of hydrolysis are compared. In particular, the 3',5'-cyclic nucleotides that have been studied have been shown to release 7 to 11 kcal mole$^{-1}$ more heat on hydrolysis than does trimethylene phosphate. Thus, despite their hydrolytic stability, 3',5'-cyclic nucleotides exhibit greater heats of hydrolysis even than do the five-membered cyclic phosphates, which are enormously labile to hydrolysis. It is by no means clear at this point why to 3',5'-cyclic nucleotides hydrolyze slowly and yet must be highly strained, nor is the origin of the strain at all obvious. Furthermore, despite the accumulation of evidence supporting the postulation of pentacoordinate trigonal-bipyramidal intermediates in the hydrolyses of five-membered cyclic phosphates, hither to unexplained factors must be present in the hydrolysis of at least the six-membered cyclic phosphates.

## 4  CYCLIC ESTERS OF SULFURIC AND SULFONIC ACIDS

The original motivation for the investigation of the hydrolysis of cyclic sulfate and sulfonate esters was provided by observations of the extreme lability of the five-membered cyclic phosphate esters and by the question of whether similar lability could be seen in ring systems other than the phosphorus series. The results obtained for the hydrolysis of the cyclic sulfates and sulfonates discussed in this review are summarized in Table 1. As the simplest model, the hydrolysis of ethylene sulfate, a five-membered cyclic sulfate ester, was investigated and compared to that of the corresponding acyclic diester dimethyl sulfate [8]. In contrast to the observations made with the five-membered cyclic phosphate esters, no exceptional lability was seen in the hydrolysis of ethylene sulfate either in neutral or alkaline solutions. The hydrolysis of ethylene sulfate proceeded only 20 times as fast as that of dimethyl sulfate in basic solution and only 100 times as fast as the hydrolysis of the corresponding six-membered sulfate ester trimethylene sulfate at 25°C. However two important points of difference were noted between five-membered cyclic sulfate esters and the acyclic and/or six-membered cyclic species. In

TABLE 1  RATE DATA FOR HYDROXIDE-ION-CATALYZED HYDROLYSIS OF SELECTED CYCLIC SULFATE AND SULFONATE ESTERS AND OF ACYCLIC ANALOGS

| Ester | $k_{OH^-}$ $(M^{-1} sec^{-1})$ | Temperature (°C) | Comments | Reference |
|---|---|---|---|---|
| Catechol cyclic sulfate | 18.8 | 25.0 | | 22 |
| Biphenylene cyclic sulfate | $1.2 \times 10^{-3}$ | 50.0 | 21% DMSO was present | 31 |
| Diphenyl sulfate | $8.9 \times 10^{-7}$ | 25.0 | Extrapolated from measurements at 50.0, 60.0 and 80.0° | 22 |
| 2-Hydroxy-5-nitro-α-toluenesulfonic acid sultone | $1.4 \times 10^3$ | 25.0 | 0.5 $M$ NaClO₄, $\mu = 0.5$ | 30 |
| 2-Hydroxy-5-bromo-α-toluenesulfonic acid sultone | 95.1 | 25.0 | 0.5 $M$ NaClO₄, $\mu = 0.5$ | 30 |
| 2-Hydroxy-α-toluenesulfonic acid sultone | 37.4 | 25.0 | 0.5 $M$ NaClO₄, $\mu = 0.5$ | 30 |
| 2-Hydroxy-5-methyl-α-toluenesulfonic acid sultone | 24.0 | 25.0 | 0.5 $M$ NaClO₄, $\mu = 0.5$ | 30 |
| 2-Hydroxy-5-methoxy-α-toluenesulfonic acid sultone | 13.6 | 25.0 | 0.5 $M$ NaClO₄, $\mu = 0.5$ | 30 |
| β-2-Hydroxyphenylethanesulfonic acid sultone | $2.8 \times 10^{-3}$ | 25.0 | | 31 |
| γ-2-Hydroxyphenylethanesulfonic acid sultone | $1.6 \times 10^{-4}$ | 50.0 | | 31 |
| Phenyl α-toluenesulfonate | $4.9 \times 10^{-5}$ | 25.0 | Extrapolated to 0% 1,2-dimethoxyethane (DME) from measurements in solutions containing DME | 26 |

particular, the enthalpy of the hydrolysis of ethylene sulfate was approximately 5 to 6 kcal mole$^{-1}$ greater than that of dimethyl sulfate [8]. Furthermore, although only carbon-oxygen bond cleavage was observed within the limits of detection in either the neutral or alkaline hydrolysis of dimethyl sulfate and trimethylene sulfate, approximately 14% sulfur-oxygen bond cleavage was seen in the reactions of ethylene sulfate with hydroxide ions. With these observations, it still remains conceivable that attack at the sulfur atom of the five-membered cyclic sulfate ring might, in fact, be greatly accelerated as compared to attack at the sulfur atom in the acyclic and six-membered cyclic sulfate esters. Also, it seemed possible that strain was present in the five-membered cyclic sulfate ring that was not relieved upon attack occurring with predominantly C—O bond fission. Thus the possibility was envisaged that if a five-membered cyclic sulfate ring were studied in which the attack of hydroxide ion was directed predominantly to the sulfur atom and if such a system exhibited ring strain similar to that present in ethylene sulfate, a large rate acceleration might well be observed when the reaction was compared to that of an acyclic ester, for example. This expectation was fulfilled when the alkaline hydrolysis of the aromatic five-membered cyclic sulfate ester catechol cyclic sulfate (I) was studied (eq. 3) and compared to that of the corresponding aromatic acyclic sulfate diester diphenyl sulfate [22]. The likelihood of attack by hydroxide ion to give aryl-oxygen cleavage in catechol cyclic sulfate was considered to be small, and thus it was expected that the rate of alkaline hydrolysis of this ester would reflect the susceptibility of the five-membered sulfate ring to attack at the sulfur atom. It was found that catechol cyclic sulfate hydrolyzes more than $10^7$ times as fast as diphenyl sulfate in alkaline solutions [22]. Furthermore, the expectation that bond cleavage in the catechol cyclic sulfate system occurs with attack of hydroxide ion at sulfur was confirmed by studies of this reaction in an $^{18}$O-enriched aqueous medium. When I was hydrolyzed in an alkaline $^{18}$O-enriched solvent and then the catechol monosulfate (II) formed was hydrolyzed further in an acidic $^{18}$O-enriched solution, the catechol isolated was found not to have excess oxygen-18. From this experiment it was concluded that neither in the hydroxide ion-catalyzed hydrolysis of catechol cyclic sulfate to give catechol mono-sulfate nor in the further acidic hydrolysis of the latter compound to give catechol and inorganic sulfate was aryl-oxygen bond fission taking place [23]. While the enthalpy of the hydrolysis of the aromatic five-membered cyclic sulfate I has not been determined, the x-ray structure determination on this diester has shown that there is a very small internal O—S—O bond angle of $97.1°$, [24] reminiscent of the small internal O—P—O bond angles seen in the five-membered cyclic phosphates. The small internal O—S—O bond angle is a common feature to all of the five-membered cyclic sulfate structures that have been determined to date. Thus in ethylene sulfate this bond angle is 98.4°, and it is $93.6.°$ in vinylene sulfate, an unsaturated cyclic sulfate system [25].

$$\text{(3)}$$

The large rate acceleration found for the reaction of an aromatic five-membered cyclic sulfate ester with hydroxide ion immediately raised the question of whether similar rate accelerations occur in related sulfur-containing esters. Therefore the reactivity of the corresponding five-membered cyclic sulfonate ester o-hydroxy-α-toluenesulfonic acid sultone (III) was examined. An approximately $10^6$-fold rate acceleration was seen for the reaction of the five-membered cyclic sulfonate with hydroxide ion as compared to that of phenyl α-toluenesulfonate [26]. This observation showed that rate accelerations in five-membered cyclic esters in the sulfur series were not confined to the cyclic sulfate systems. It is interesting to note that the six-membered cyclic sulfonate ester β-o-hydroxy-phenylethanesulfonic acid sultone (IV) is hydrolyzed somewhat more rapidly in base than phenyl α-toluenesulfonate, but still at a rate that is very much diminished as compared to the five-membered cyclic species. Once again the x-ray structure determination of the five-membered cyclic ester indicated that the internal bond angle at the sulfur-containing group was small. The C—S—O bond angle was only 96.1°. This is in contrast to to the corresponding angle in the six-membered cyclic sulfonate (IV) which is 101.4°, or 5.3° larger.

While the rate acceleration observed for the hydrolysis of catechol cyclic sulfate relative to diphenyl sulfate can be ascribed to the differences in the rates of attack of hydroxide ion at the sulfur atoms in a five-membered cyclic sulfonate and an open-chain sulfate, respectively, in the case of the five-membered sulfonate **III** hydrolytic mechanisms that do not involve direct attack of hydroxide ion at sulfur could give rise to sulfur-oxygen bond fission. Two such mechanisms are given by eq. 4 and 5.

$$(4)$$

$$(5)$$

The mechanisms shown in eq. 4 and 5 involve the postulated intermediacy of carbanions and/or sulfenes in the hydrolysis of the cyclic sulfonate **III**. Through experiments carried out in a $D_2O$-$OD^-$ solution where the sultone **III** was in excess over $OD^-$ and **III** could be recovered after all the $OD^-$ was consumed, it was found that **III** had undergone considerable exchange of deuterium in the methylene group. The observation that a carbanion is formed rapidly and reversibly from the sultone **III** in basic solution allows the exclusion of the concerted pathway of eq. 4 as the principal route in the alkaline hydrolysis of o-hydroxy-α-toluenesulfonic acid sultone. However the evaluation of the importance of the pathway of eq. 5 is a far more difficult matter.

As is discussed later in this chapter, in the alkaline hydrolysis of 5-nitrocoumaranone, a five-membered lactone with labile α-protons that undergo ionization with a p$K_a$ of 9.8, an experimental approach involving the measurement of the solvent isotope effect for the decomposition of the fully ionized ester was used to exclude a mechanism analogous to that given in eq. 5 [27]. However the method used is limited to cases in which the carbon acid is fully ionized at an alkalinity accessible in aqueous media. To measure the $pK_a$ for the ionization of the labile $\alpha$-protons of $o$-hydroxy-$\alpha$-toluene-sulfonic acid sultone (III), it has been necessary to use a nonaqueous medium. In dimethyl sulfoxide a $pK_a$ value of 15.9 has been obtained for the ionization of an $\alpha$-proton of the methylene group of III [28]. Despite the impossibility of studying sultone III in an aqueous medium at $25°C$ under conditions where full ionization at the methylene position can be attained, a number of general methods have been developed involving comparative isotope exchange and hydrolysis rate measurements to show that mechanisms like that of eq. 5, which involve carbanion and/or sulfene intermediates, do not provide the predominant pathways by which the five-membered cyclic sulfonate III hydrolyzes. Thus the large rate enhancements observed for the alkaline hydrolysis of the five-membered cyclic ester III relative to its open-chain analog and the six-membered cyclic species IV almost certainly reflect the differences in the rate of attack of hydroxide ion at the sulfur atoms between the five-membered cyclic and the acyclic or six-membered cyclic systems.

It is noteworthy that recently evidence has been provided that an elimination-addition route including sulfene formation, like that of eq. 5, is involved in the hydrolysis of phenyl $\alpha$-toluenesulfonate. Therefore the actual difference in the rates of attack of hydroxide ion at the sulfur atoms in III and in phenyl $\alpha$-toluenesulfonate must be significantly greater than the observed factor of $10^6$ [29].

In view of the importance of pentacoordinate intermediates in the hydrolysis of five-membered cyclic phosphates and the various similarities between these compounds and the five-membered cyclic sulfates and sulfonates, a question that naturally arises is whether there might be reversible formation of pentacoordinate intermediates during the hydrolysis of the sulfur-containing esters. To explore this question, incomplete hydrolyses of catechol cyclic sulfate, $o$-hydroxy-$\alpha$-toluenesulfonic acid sultone (III), and $\beta$-$o$-hydroxyphenyl-ethanesulfonic acid sultone (IV), were carried out in alkaline solutions containing excess oxygen-18 [23]. When the unconverted starting esters were reisolated, no significant exchange was found, showing that in the hydrolysis reactions there is no detectable reversible formation of pentacoordinate intermediates such as V, VI, or VII, in which the oxygens attached to the sulfur atom and external to the ring have been equilibrated. However it should be pointed out that the possibility cannot be ruled out that pentacoordinate intermediates are formed irreversibly or that pentacoordinate intermediates might be formed reversibly in the hydrolyses of the cyclic esters, and the oxygens external to the ring might not equilibrate during the lifetimes of such intermediates.

V                           VI                          VII

VIII

In contrast to the situation seen for the cyclic phosphates, acid catalysis has not been observed in the hydrolysis of cyclic sulfates and sulfonates. The primary evidence for the formation of the penta-coordinate intermediates in the hydrolysis of five-membered cyclic phosphate diesters like ethylene phosphate has come from experiments done in acidic oxygen-18 enriched solutions, since the alkaline hydrolysis of ethylene phosphate is not accompanied by oxygen exchange [5]. A possible explanation for the latter fact is that the pentacoordinate intermediate VIII, which might be formed in alkaline solution, would have a double negative charge and could conceivably undergo ring cleavage or loss of the attacking hydroxide ion faster than the proton shift needed to achieve oxygen exchange.

With regard to the cyclic sulfur-containing esters, if one postulates that pentacoordinate intermediates with trigonal-bipyramidal geometry are formed in the hydrolysis reactions and that the analysis that has been applied to the problem of pseudorotation in the hydrolysis of phosphate esters [3] can be applied in the sulfur series, then the absence of oxygen exchange in the alkaline hydrolysis of compounds like I, III, and IV can be readily understood. For example, if one considers possible trigonal-bipyramidal geometries for V, the species that might be an intermediate in the hydroxide-ion-catalyzed hydrolysis of catechol cyclic sulfate, then structure IX, in which the sulfur-containing ring occupies one apical and one equatorial position, the internal O—S—O bond angle is 90°, and the two negatively charged groups are in equatorial positions, should be favored substantially over structure X in which one of the negatively charged groups is placed in an apical position. This means that pseudorotation from structure IX to structure X should be unlikely and that if oxygen exchange involves such pseudorotation, it may not be competitive with ring cleavage. Of course, in principle this difficulty could be circumvented if proton transfer to the incipient apical oxygen would occur concertedly with pseudorotation between IX and X.

IX                                    X

As a possible probe of the transition states in the hydrolyses of five-membered cyclic sulfur-containing esters, a classical Hammett study of substituent effects was performed with a series of 5-substituted 2-hydroxy-α-toluene sulfonic acid sultones (XI) [30]. A linear relationship was found between the logarithms of the rate constants

XI

Y = H, NH$_2$, OCH$_3$, CH$_3$, Br, NO$_2$

for the alkaline hydrolysis of these compounds and the appropriate Hammett para substituent constants, $\sigma_p$. A positive $\rho$ value of +1.23 was obtained, indicating that electron-withdrawing substituents in the 5-position of the aromatic ring have an accelerating effect on the alkaline hydrolysis of aromatic five-membered cyclic sulfonates. If the mechanism of hydrolysis of the sultones involves a concerted nucleophilic displacement reaction by hydroxide ion at sulfur, then these observations suggest that the sulfur-oxygen bond is significantly cleaved in the transition states for hydrolysis. In the event, however, that pentacoordinate intermediates do lie along the pathway for the alkaline hydrolysis of sultones, the interpretation of the $\rho$ value found becomes somewhat ambiguous. Nevertheless, it seems likely that the $\rho$ value for the first step in the hydrolysis mechanism given in eq. 6 would not be large and that the observed $\rho$ value would reflect primarily the effects of substituents on the reaction of step 2 in which the pentacoordinate intermediate undergoes ring opening. According to this hypothesis, the transition state for the ring-opening reaction (step 2) would be quite polar.

$$(6)$$

An interesting possibility that could arise if pentacoordinate intermediates are involved in the hydrolysis of cyclic sulfates and sulfonates is that species like **V**, **VI**, and **VII** might ionize in basic media to give the corresponding dinegatively charged species, such as **XII**. If this occurred the rate law for the alkaline hydrolysis of the cyclic esters could contain a kinetic term second order in hydroxide ion concentration. To test this possibility the kinetics of the hydrolysis of the relatively stable sultones **IV** and **XIII** and the cyclic sulfate **XIV** were studied in strongly basic media [31]. The rate constants obtained showed a first order dependence on the function $a_w 10^{H-}$, indicating that the hydrolysis reactions were first order in hydroxide ion [32]. While these results certainly do not rule out the possibility that pentacoordinate intermediates in which the attacking hydroxide ion is covalently bound to sulfur are formed in the hydrolysis of **IV**, **XIII**, and **XI** under basic conditions, they do not require the postulation of such species.

In summary, there is no compelling evidence at the present time for the postulation of pentacoordinate intermediates in the alkaline hydrolysis of the cyclic sulfates or sulfonates. However the transition states in the hydrolytic reactions of the highly reactive five-membered cyclic esters may have structures with approximately trigonal-bipyramidal geometry in which the ring angle at sulfur is close to $90°$ and the five-membered ring spans one apical and one equatorial position.

The need for relatively little perturbation of the ring angle at sulfur in the five-membered cyclic esters to achieve such a transition state geometry would be consistent with the high degree of hydrolytic lability of these compounds.

## 5 HYDROLYSIS OF LACTONES

The first systematic study of the influence of the ring size on the rate of alkaline hydrolysis of aliphatic lactones was carried out by Huisgen and Ott [33]. The essentials of their kinetic results are summarized in Table 2, along with the molecular dipole moments of

TABLE 2  BASE-CATALYZED HYDROLYSIS OF 4- TO 16-MEMBER
RING LACTONES AND COMPARABLE ESTERS IN
60:40 DIOXANE/WATER [33]

| Ring Size | $\mu_{(D)}$ | Configuration | $10^4 \times k_{OH^-} (M^{-1} sec^{-1})$ 0 C |
|---|---|---|---|
| $4^a$ | | | $12,000^a$ [34] |
| 5 | 4.09 | cis | 1,480 |
| 6 | 4.22 | cis | 55,000 |
| 7 | 4.45 | cis | 2.500 |
| 8 | 3.70 | cis, trans | 3,500 |
| 9 | 2.25 | cis, trans | 116 |
| 10 | 2.01 | trans | 0.22 |
| 12 | | trans | 3.3 |
| 14 | 1.86 | trans | 3.00 |
| 16 | | trans | 6.5 |
| n-Butyl caproate | 1.79 | trans | 8.4 |

$^a$In water.

the lactones, reflecting the transition from cis to trans esters in the region of eight- to nine-member rings. Comparison of the rate constants of lactone hydrolysis with that of the arbitrary reference compound n-butyl caproate shows that five- seven-, and eight-member rings are hydrolyzed about 200 to 300 times as fast as open-chain esters, whereas the larger rings display no rate-accelerating effects. The four- and six-membered rings are hydrolyzed at even larger rates, by factors of 1200 and 5500, respectively. The large rate accelerations observed with small ring lactones could be attributed *a priori* to a variety of factors. Steric strain, restriction of the ester to the cis configuration, distortion of the normally coplanar ester function, and

## TABLE 3 RATE CONSTANTS PERTINENT TO LACTONE HYDROLYSES

| | Compound | $k_{OH^-} (M^{-1} sec^{-1})$ | Conditions | Reference |
|---|---|---|---|---|
| XV | 5-Nitro-2-coumaranone | 1290 | 0.2 $M$ KCl, 1.6% CH$_3$CN, 25°C | 34 |
| XVI | 2-Coumaranone | 85 (if p$K_a$ = 12.25) | 0.2 $M$ KCl, 1.6% CH$_3$CN, 25°C | 34 |
| XVII | 6-Nitro-3,4-dihydrocoumarin | 2740 | 0.2 $M$ KCl, 1.6% CH$_3$CN, 25°C | 34 |
| XVIII | 3,4-Dihydrocoumarin | 824 | 0.2 $M$ KCl, 1.6% CH$_3$CN, 25°C | 34 |
| XIX | $p$-Nitrophenyl acetate | 9.4 | 0.2 $M$ KCl, 1.6% CH$_3$CN, 25°C | 34 |
| XX | $p$-Nitrophenyl $\beta$-phenylpropionate | 14.6 | 0.2 $M$ KCl, 1.6% CH$_3$CN, 25°C | 34 |
| XXI | $p$-Nitrophenylphenylacetate | 62 | 0.2 $M$ KCl, 1.6% CH$_3$CN, 25°C | 34 |
| XXII | $p$-Nitrophenyl $m$-nitrophenylacetate | 110 | 0.2 $M$ KCl, 1.6% CH$_3$CN, 25°C | 34 |
| XXIIA | Phenyl acetate | 1.3 | 1 $M$ KCl, 25°C | 35 |
| XXIII | $n$-Butyl caproate | 0.00084 | 60% Dioxane/H$_2$O, 0°C | 33 |
| XXIV | $\gamma$-Butyrolactone | 0.148 | 60% Dioxane/H$_2$O, 0°C | 33 |
| XXV | $\delta$-Valerolactone | 5.5 | 60% Dioxane/H$_2$O, 0°C | 33 |
| XXVI | $\epsilon$-Heptyllactone | 0.25 | 60% Dioxane/H$_2$O, 0°C | 33 |
| XXVII | $\omega$-Octyllactone | 0.35 | 60% Dioxane/H$_2$O, 0°C | 33 |
| XXVIII | Phthalide | 0.144 | 33% Dioxane/H$_2$O, 25°C | 36 |
| XXIX | Benzyl benzoate | 0.01 | (Calculated from 60% Dioxane/H$_2$O)25°C | 36 |
| XXX | Cyclopentanone and semicarbazide | $k$=0.153 $M^{-1}$ sec$^{-1}$ | 25°C | 37 |
| XXXI | Cyclohexanone and semicarbazide | $k$=0.152 $M^{-1}$ sec$^{-1}$ | 25°C | 37 |

change in the mechanism or in the rate limiting step alone or in combination are all possible causes. That ring size by itself is not the factor responsible for the acceleration of the ring-opening reaction is demonstrated by the example of phthalide (Table 3), which hydrolyzes no more than 10 times as fast as its open-chain equivalent benzyl benzoate. By virtue of the very same example, the restriction of the ester configuration into a cis structure could not account alone for the large effects observed with rings of four to eight members.

In the following discussion, the contribution of each of the possible rate-accelerating factors is examined in relation to the structure of the ester-containing ring. In Table 3 we summarize most of the kinetic data available for lactone hydrolysis. We hope to show that indeed each of the accelerating factors contributes in general to the lability of the lactone ring and that none of them is so powerful as to be singled out as a hitherto unrecognized catalytic panacea.

## 5.1   Change in Mechanism

The alkaline hydrolysis of lactones is in general first order with respect to hydroxide ion and is thus kinetically indistinguishable from the alkaline hydrolysis of esters, which has been shown to occur with an addition-elimination mechanism. However first order dependency on hydroxide ion could also be consistent with an elimination-addition pathway, if the reactive species were a carbanion. It has indeed been observed that esters and amides possessing a readily ionizable proton adjacent to the carboxyl group are hydrolyzed in alkaline solution by way of a pathway involving an isocyanate or a ketene intermediate according to the general scheme:

$$HX - \overset{\overset{O}{\diagup\diagdown}}{C}\!\!-\!\!OR \underset{}{\overset{K_a}{\rightleftharpoons}} H^{\oplus} + {}^{\ominus}X\!-\!\overset{\overset{O}{\diagup\diagdown}}{C}\!\!-\!\!OR \overset{k_1}{\longrightarrow} {}^{\ominus}OR + X\!=\!C\!=\!O$$

$$HX\!-\!\overset{\overset{O}{\diagup\diagdown}}{C}\!\!-\!OH \overset{\text{fast}}{\underset{+H_2O}{\longleftarrow}} \qquad (7)$$

where XH represents the carbon-acid or nitrogen-acid portion of the molecule. Even if the ionized species ${}^{\ominus}X\!-\!CO\!-\!OR$ is detected in the reaction mixture, it is difficult to establish beyond doubt its intermediacy in the reaction pathway, since eq. 7 is kinetically equivalent to a pathway where the ionized substrate is a mere nonreactive byproduct, as shown in eq. 8.

$$H^{\oplus} + {}^{\ominus}X-C{\overset{O}{{\diagdown}}}-R \underset{}{\overset{K_a}{\rightleftharpoons}} HX-C{\overset{O}{{\diagdown}}}-OR \underset{+OH^{\oplus}}{\overset{k_2}{\longrightarrow}} HX-C{\overset{O}{{\diagdown}}}-OH + RO^{\ominus} \quad (8)$$

The ionization of the α-protons of lactones and the possibility of ketene intermediates in lactone hydrolysis was investigated with the aromatic nitro-substituted lactone 5-nitrocoumaranone (**XV**), a compound readily amenable to analysis by spectrophotometry [34].

In the alkaline solution the hydrolysis of **XV** is itself a biphasic reaction: at 400 nm a rapid burst of absorbance is followed by a much slower formation of the final acid product. At 500 nm the rapid formation and slow decomposition of an intermediate is observed.

By analysis of the reaction kinetics and the pH dependency of the spectral properties of the intermediate, it was shown that it is indeed the carbanion resulting from the ionization of the α-proton (eq. 9). The formation of the carbanion was also demonstrated for

$$(9)$$

**XV**

2-coumaranone (**XVI**). The equilibrium constants for the ionization of these carbon-acids at 25°C are summarized in Table 4.

**TABLE 4   THE IONIZATION OF LACTONES AND THEIR ANALOGS [34]**

| Carbon-Acid | pK |
|---|---|
| 5-Nitro-2-indanone | 8.95 (9.48 in $D_2O$) |
| 5-Nitro-2-coumaranone | 9.8 |
| 2-Coumaranone | 12-12.5 |
| p-Nitrophenyl p-nitrophenylacetate | >14 |

Comparison of these ionization constants with those for nitro-substituted phenols [38] suggests that substituent effects are similar in the two series of acids. In the carbon-acids, substitution of the para hydrogen by a nitro group decreases the pK by 3 to 3.5 units. The analogous phenols differ by 2.8 units in the same direction, thereby indicating a ρ of the same magnitude for the two ionization processes. In the same manner the *meta*-nitro and the *para*-nitro carbon- and oxygen-acids differ by 0.9 and 1.1 pH units, respectively. Deuterium oxide solvent isotope effects on the two systems are again similar, the effect on these carbon-acids is the raise the p$K$ by 0.5 units, while the average value for oxygen-acids is 0.55 p$K$ units.

The rate constants for the ionization process of **XV** are defined by eqs. 10 and 11. The experimentally obtained values of the rate

$$H_2O + NC \underset{k_b}{\overset{k_a}{\rightleftharpoons}} NC^- + H_3O^+ \tag{10}$$

$$OH^- + NC \underset{k_d}{\overset{k_c}{\rightleftharpoons}} NC^- + H_2O \tag{11}$$

constants are given in Table 5 together with those for acetylactone [39].

TABLE 5   IONIZATION OF XV AND ACETYLACETONE AT 25°C [34,39]

| Compound | $k_a(H_2O)$ (sec$^{-1}$) | $k_b$ ($M^{-1}$sec$^{-1}$) | $k_c$ ($M^{-1}$sec$^{-1}$) | $k_d(H_2O)$ (sec$^{-1}$) | p$K$ |
|---|---|---|---|---|---|
| 5-Nitro-2-coumaranone | 0.155 | $9.7 \times 10^8$ | $4.9 \times 10^4$ | 2.5 | 9.8 |
| Acetylacetone | 0.014 | $0.12 \times 10^8$ | $4 \times 10^4$ | 0.35 | 8.9 |

These results establish that in alkaline solutions the five-membered aromatic lactones are completely ionized and that the acidity of the α-proton is considerably enhanced by ring closure.

The low p$K$'s observed for these compounds relative to their open-chain ester analog **XXII** may be reasonably ascribed to two effects. First, one of the canonical structures of the carbanion of **XV**, which contributes to the stability of the ion, is analogous to the structure of benzofurane. This additional aromatic character is, of course, not available in the carbanion derived from **XXII**. Second, while the carbanion derived from 5-nitro-2-indanone cannot be resonance stabilized by formation of a new aromatic system, the carbonyl system is more strongly electron withdrawing in ketones than in esters. This then would result in a larger conjugation with the benzene ring in 5-nitro-2-indanone than in **XXII**.

The measurement of the kinetic solvent isotope effects demonstrated that the increased lability of the α-hydrogen of lactones with respect to aliphatic esters is not accompanied by a change from the addition-elimination to the elimination-addition mechanism [40]. If the carbanion is a rapid equilibrium with the lactone, then the experimentally observed first order rate constant ($k_{exp}$) is given by

$$k_{exp} = \frac{k_1}{1 + [H]/K_a} \tag{12}$$

eq. 12 for the ketene pathway (eq. 7) and by eq. 13 for the hydro-

$$k_{exp} = \frac{k_2 K_w}{K_a + [H]} \tag{13}$$

lysis through a tetrahedral intermediate (eq. 8).

Equation 12 states that in strongly alkaline solution ($pH \gg pK_a$) $k_{exp}$ is equal to $k_1$. Since the elimination step of eq. 7 does not involve a proton transfer $k_{exp}^{D_2O}/k_{exp}^{H_2O} = 1$. Under the same conditions, eq. 13 yields

$$\frac{k_{exp}^{D_2O}}{k_{exp}^{H_2O}} = \frac{k_2^{D_2O}}{k_2^{H_2O}} \quad \frac{K_w^{D_2O}}{K_w^{H_2O}} \quad \frac{K_a^{D_2O}}{K_a^{H_2O}}.$$

Using values of $k^{D_2O}/k^{H_2O} = 1.22$, $K_w^{D_2O}/K_w^{H_2O} = 0.15$, and $K_a^{H_2O}/K_a^{D_2O} = 3.5$, we estimate that $k_{exp}^{D_2O}/k_{exp}^{H_2O}$ should be approximately 0.65.

Experimentally the value of $k_{exp}^{D_2O}/k_{exp}^{H_2O}$ has been found to be 0.58 for the alkaline hydrolysis of XV, thus excluding any major contribution of the elimination mechanism to lactone hydrolysis. Since some ionizable trans esters do hydrolyze via an elimination-addition pathway [41], the question arises then of why cyclic esters are different in their behavior. It is suggested that even lactones form a ketene but that this step is rapidly reversible and that the ketene exclusively yields the original carbanion by a kinetic control rather than the carboxylic acid by thermodynamic control.

## 5.2 Change in Rate-Limiting Step

Evidence for the existence of a tetrahedral intermediate in the hydrolysis of esters comes from [18]O-exchange experiments [42]. The ester labeled in the carbonyl oxygen is partially hydrolyzed in ordinary water and the [18]O remaining in the carbonyl oxygen is determined after isolation of the unreacted ester (see Table 6). The rate

## TABLE 6 OXYGEN EXCHANGE ACCOMPANYING THE ALKALINE HYDROLYSIS OF BENZOATE ESTERS AND LACTONES [43, 44]

| Compound | $k_h/k_{ex}$ |
|---|---|
| Benzoates | |
| Methyl | 5.8 |
| Ethyl | 10.6 |
| Phenyl | >105 |
| $p$-Cl-benzyl | > 60 |
| $p$-MeO-benzyl | >192 |
| Phthalide | > 50 |
| $\gamma$-Butyrolactone | > 30 |

$$
\begin{array}{c}
\overset{^{18}O}{\overset{\|}{R'-C-OR}} + H_2{}^{16}O \\[2em]
\overset{^{16}O}{\overset{\|}{R'-C-OR}} + H_2{}^{18}O
\end{array}
\;\underset{k_2}{\overset{k_1}{\rightleftharpoons}}\;
\left[\begin{array}{c}
^{18}OH \\ | \\ R'-C-OR \\ | \\ ^{16}OH
\end{array}\right]
\;\xrightarrow{k_3}\; R'COOH + HOR
$$

constant for the hydrolytic reaction is $k_h = (k_1/k_3)/(k_2 + k_3)$ and if $k_{ex}$ is the rate constant of exchange of $^{18}O$, then $k_h/k_{ex} = 2k_3/k_2$. If $k_h/k_{ex} \gg 1$, then $k_3 \gg k_2$ and $k_h = k_1$. In other words, if $k_h/k_{ex} \gg 1$, then $k_h$ measures the rate of formation of the tetrahedral intermediate whereas if $k_h/k_{ex} \simeq 1$, $k_h$ is a combination of rate constants. Values of $k_h/k_{ex}$ determined for a variety of compounds (Table VI) lead to the following conclusion: in the hydrolysis of all the compounds with good leaving groups the rate of hydrolysis reflects the rate of addition of the nucleophile to the carbonyl carbon. The rate of hydrolyiss of **XXIII**, however, reflects all three rate constants, $k_1$, $k_2$, and $k_3$, defined above. Thus comparison of the rates of hydrolysis of **XXIV** and **XXIII** does not compare identical processes, and their ratio (176) reflects as well the acceleration due to the elimination step as the possible acceleration in the addition step.

## 5.3 Steric Strains

The importance of the stereochemistry of the ester and that of the tetrahedral intermediate in determining the rate or ring opening was recognized by Brown, et al. [36]. These authors reasoned that in the case of six-membered lactones, the trigonal carboxyl carbon unfavorably disturbs the perfectly staggered chair conformation of the cyclohexane ring by transforming it into a half chair with an exo double bond and four eclipsed H atoms. The formation of the tetrahedral intermediate restores the unhindered chair conformation and is thus accompanied by a considerable relief of steric strains. In the case of the roughly planar, five-membered δ-valerolactone, on the other hand, the formation of the tetrahedral intermediate introduces additional bond oppositions, since the ring remains slightly puckered and is thus sterically unfavorable. For this reason the larger rate of hydrolysis of six-membered rings in comparison with five-membered ones can be accounted for by development or relief of steric strains in the addition-elimination mechanism.

Using the hypothesis that a tetrahedral addition complex relieves the strain in six-membered rings containing a trigonal carbon, but not in five-membered rings, Brown et al. [36] were able to rationalize a host of kinetic and thermodynamic data related to saturated five- and six-membered ring systems. The steric strain factor was found to play an important role in a wide variety of reactions, such as equilibria involving exo and endo double bond isomers and tautomers, olefin formation in terpenes and related compounds, the stability of lactones, hemiacetals, imides, and cyclic carbonate esters, and the formation of furanose and pyranose rings of sugars and sugar acids. Since in cyclic carbonates the trigonal carbon is located between two ether oxygens, the presence of a steric strain factor in their hydrolysis indicates that the nonbonding electrons of the ether oxygen have spatial requirements similar to those of hydrogen atoms in methylene groups.

The data on the alkaline hydrolysis of lactones, reported in Table 3, also show convincingly the presence of the steric-strain effect. The comparison of the rate constant for five-membered rings and six-membered rings is shown in Table 7. After appropriate corrections for substituent effects, the magnitude of the steric-strain effect is of the order of 15- to 40-fold and it appears to be independent of the presence of an unsaturated linkage of the ring. In comparison, the addition of semicarbazide to ketones shows a quite similar, 10-fold effect.

Small-ring aliphatic lactones are hydrolyzed at very similar rates, with the exception of valerolactone (Table 3). This fact would argue in favor of the steric strain being operative exclusively in six-membered rings. Apparently the geometry of smaller or larger ring sizes is not conducive to relief of the strain in the tetrahedral addition intermediate.

# TABLE 7  STERIC HINDRANCE EFFECTS—ADDITION TO EXO DOUBLE BONDS IN FIVE- AND SIX-MEMBERED RINGS

| | Rate Constant | | Effect = $\dfrac{\text{Six-Member Rate}}{\text{Five-Member Rate}}$ |
|---|---|---|---|
| | Five-Member Ring ($n = 3$) | Six-Member Ring ($n = 4$) | |
| $(CH_2)_n C \begin{smallmatrix}=O\\\\O\end{smallmatrix}$ + OH⁻ | 0.148 | 5.5 | 37 |
| NO₂— (benzo) $(CH_2)_{n-2}$ C=O + OH⁻ | 1290 | 2740 | 15.8[a] |
| (benzo) $(CH_2)_{n-2}$ C=O + OH⁻ | 85 | 824 | 41.2[a] |
| $(CH_2)_n C \begin{smallmatrix}=O\\\\CH_2\end{smallmatrix}$ + semicarbazide | 0.0153 | 0.152 | 10.0 |

[a] After correction for substituent effects calculated from hydrolysis of **XXI**, **XX**, and **XXII**.

## TABLE 8 RING STRAIN AND CIS EFFECT—HYDROLYSIS OF ANALOGOUS TRANS ESTERS AND FIVE-MEMBERED CIS LACTONES

| Cis Compound | $k_{OH^-} (M^{-1} sec^{-1})$ | Trans Compound | $k_{OH^-} (M^{-1} sec^{-1})$ | Ratio $= \dfrac{k\,(\text{cis})}{k\,(\text{trans})}$ |
|---|---|---|---|---|
| $\gamma$-Butyrolactone | 0.148 | n-Butyl caproate | 0.00084 | 176 |
| 2-Coumaranone | 85 | Phenyl acetate | 1.3 | 65 |
| Phthalide | 0.144 | Benzyl benzoate | 0.01 | 14.4 |
| 5-Nitro-2-coumaranone | 1290 | p-Nitrophenyl p-nitro-phenyl acetate | 110 | 11.7 |

## 5.4   Ring Strain and cis Effect

If indeed the alkaline hydrolysis of five-membered lactones is free of the steric-strain factor, then it can be used to assess the net contribution of ring strain and of cis-ester configuration to the increase in rate. Table 8 shows a comparison of the alkaline rate constants of five-membered lactones with that of their corresponding open-chain analogs. **XXVIII** and **XV** display a 10 to 15-fold increase in rate in comparison with the corresponding noncyclic esters. This modest accelerating effect is attributable to the increased lability of cis esters with respect to that of trans esters, since ring strain is not expected to be important in **XXVIII** and **XV**. Alternatively, at least part of this factor could arise from the increased accessibility of the carbonyl groups in a rigid ring in comparison with the shielding effect exerted by alkyl chains in random motion around it.

A somewhat larger factor is observed with the five-membered rings of **XVI** and of **XXIV**, which after correction for the cis effect amounts to 5- to 12-fold. In the light of the considerations involving the rate of formation and decomposition of the tetrahedral intermediate, we feel that this accelerating effect can be correlated with the rather poor leaving groups in both esters. Thus the open-chain reference compounds are probably hydrolyzed with a rate where addition is not solely rate limiting. On the other hand, due to the favorable entropy of activation of ring opening in the elimination step, the limiting step in the hydrolysis of the corresponding lactones should be the addition of the nucleophile.

## 6   CONCLUSIONS

A general comparison of the kinetic data for the hydrolysis of cyclic carboxylic, phosphoric, sulfuric, and sulfonic acid esters of moderate ring size with the corresponding open-chain esters shows that the contribution of the ring structure is to either accelerate or at least maintain the hydrolytic reactivity of the ester function. These findings indicate tnat the ring structure does not impart any special stability to the esters. The rate accelerations seen for the cyclic esters relative to the acyclic analogs vary enormously, sometimes reaching factors exceeding a millionfold. The causes of such accelerations have been ascribed variously to mechanistic changes, ring strain, steric strain, cis-trans isomerization of the ester function, and steric accessibility of the ester group.

While each of these factors may play a role in the reactions of certain types of cyclic esters, no single factor accounts for all the observations concerning their hydrolytic lability. Thus ring strain appears to be dominant in the hydrolysis of five-membered cyclic phosphates, sulfates, and sulfonates, whereas it is a minor contributor to the reactivity of the corresponding lactones. In fact, the kinetic data suggest that the lactones are not significantly strained. Again, while mechanistic differences are not observable in comparing moderately sized lactones to open-chain esters, recent evidence suggests that despite the unimportance of an elimination-addition route in the solvolysis of five-membered cyclic sulfonates, such a route does operate in the alkaline hydrolysis of related open-chain sulfonates [29]. Indeed, in the case of lactone hydrolysis the largest rate acceleration, that seen with δ-valerolactone, is only somewhat greater than $10^3$, and it is due to a combination of minor contributing factors.

At the present time there are still many unanswered questions concerning cyclic ester hydrolysis. For example, it might seem surprising that in highly strained rings like that of o-hydroxy-α-toluenesulfonic acid sultone the elimination-addition pathway (a direct route to the relief of ring strain) is not followed. An explanation of this observation based on the unfavorably geometry of the sulfene function which would be produced by ring opening has been proposed [29], but the validity of this hypothesis remains to be tested. Furthermore, the reasons for the high degree of ring strain in the five-membered cyclic phosphates, sulfates, and sulfonates as compared to the apparent absence of ring strain in the related lactones and lactams is not completely understood. Finally, the source of the ring strain in the 3',5'-cyclic nucleotides remains to be elucidated.

Although the reactivity of the cyclic esters is itself an interesting problem, the major reason for the current interest in their study is that many biologically important molecules contain cyclic phosphate and lactone structures. Several hypotheses have been advanced establishing a correlation between the reactivity and the structure of these molecules. Thus it has been proposed that the structure of the cyclic AMP system is particularly suited to the reversible adenylylation of the effector sites in cyclic AMP-stimulated enzymes [19]. A similar suggestion has been advanced to account for the effects of the hypothalamic hormones containing the pyroglutamic acid ring system [45]. A good understanding of the importance of the cyclic ester (and amide) structure to reactivity in biological systems may aid greatly in the design of hormone analogs and drugs and in the elucidation of their mechanisms of action.

## ACKNOWLEDGMENT

The authors gratefully acknowledge grants from the National Institute of Health and the National Science Foundation.

## REFERENCES

1. J. Kumamoto, J. R. Cox, Jr., and F. H. Westheimer, *J. Am. Chem. Soc.*, 78, 4858 (1956).
2. J. R. Cox, Jr. and B. Ramsay, *Chem. Rev.*, 64, 317 (1964).
3. F. H. Westheimer, *Acc. Chem. Res.*, 1 70 (1968).
4. S. M. Brown, D. I. Magrath, and A. R. Todd, *J. Chem. Soc.*, 1952, 2708.
5. P. C. Haake and F. H. Westheimer, *J. Am. Chem. Soc.*, 83, 1102 (1961).
6. H. G. Khorana, G. M. Tener, R. S. Wright, and J. G. Moffatt, *J. Am. Chem. Soc.*, 79, 430 (1957).
7. M. Smith, G. I. Drummond, and H. G. Khorana, *J. Am. Chem. Soc.*, 83, 698 (1961).
8. E. T. Kaiser, M. Panar, and F. H. Westheimer, *J. Am. Chem. Soc.*, 85, 602 (1963).
9. P. Greengard, S. A. Rudolph, and J. M. Sturtevant, *J. Biol. Chem.*, 244, 4789 (1969); S. A. Rudolph, E. M. Johnson, and P. Greengard, *J. Biol. Chem.*, 246, 1271 (1971).
10. J. M. Sturtevant, J. A. Gerlt, and F. H. Westheimer, *J. Am. Chem. Soc.*, 95, 8168 (1973).
11. T. A. Steitz and W. N. Lipscomb, *J. Am. Chem. Soc.*, 87, 2488 (1965).
12. C. L. Coulter and M. L. Greaves, *Science*, 169, 1097 (1970); C. L. Coulter, *J. Am. Chem. Soc.*, 95, 570 (1973).
13. D. A. Usher, E. A. Dennis, and F. H. Westheimer, *J. Am. Chem. Soc.*, 87, 2320 (1965).
14. F. Covitz and F. H. Westheimer, *J. Am. Chem. Soc.*, 85, 1773 (1963).
15. E. A. Dennis and F. H. Westheimer, *J. Am. Chem. Soc.*, 88, 3431 (1966); 88, 3432 (1966).
16. D. Gorenstein and F. H. Westheimer, *J. Am. Chem. Soc.*, 89, 2762 (1967).
17. R. Kluger, F. Kerst, D. Lee, and F. H. Westheimer, *J. Am. Chem. Soc.*, 89, 3919 (1967).
18. E. T. Kaiser and K. Kudo, *J. Am. Chem. Soc.*, 89, 6725 (1967).
19. E. T. Kaiser, T. W. S. Lee, and F. P. Boer, *J. Am. Chem. Soc.*, 93, 2351 (1971).
20. M. G. Newton, J. R. Cox, Jr., and J. A. Bertrand, *J. Am. Chem. Soc.*, 88, 1503 (1966).
21. D. Swank, C. N. Caughlan, F. Ramirez, O. P. Madan, and C. P. Smith, *J. Am. Chem.* 89, 6503 (1967).
22. E. T. Kaiser, I. R. Katz, and T. F. Wulfers, *J. Am. Chem. Soc.*, 87, 3781 (1965).
23. E. T. Kaiser and O. R. Zaborsky, *J. Am. Chem. Soc.*, 90, 4626 (1968).

24. F. P. Boer and J. J. Flynn, *J. Am. Chem. Soc.*, **91**, 6604 (1969).

25. F. P. Boer, J. J. Flynn, E. T. Kaiser, O. R. Zaborsky, D. A. Tomalia, A. E. Young and Y. C. Tong, *J. Am. Chem. Soc.*, **90**, 2970 (1968).

26. O. R. Zaborsky and E. T. Kaiser, *J. Am. Chem. Soc.*, **88**, 3084 (1966).

27. P. S. Tobias and F. J. Kézdy, *J. Am. Chem. Soc.*, **91**, 5171 (1969).

28. P. Müller, D. F. Mayers and E. T. Kaiser, manuscript in preparation.

29. A. Williams, K. T. Douglas, and J. S. Loran, *J. Chem. Soc. Chem. Commun.*, 689 (1970).

30. O. R. Zaborsky and E. T. Kaiser, *J. Am. Chem. Soc.*, **92**, 860 (1970).

31. J. H. Smith, I. Inoue, and E. T. Kaiser, *J. Am. Chem. Soc.*, **94**, 3098 (1972).

32. C. H. Rochester, *Acidity Functions*, Academic Press, New York, 1970.

33. R. Huisgen and H. Ott, *Tetrahedron*, **6**, 253 (1959).

34. P. S. Tobias, Ph.D. Thesis, University of Chicago, 1971.

35. W. P. Jencks, and J. Carriuolo, *J. Am. Chem. Soc.*, **82**, 1778 (1960).

36. H. C. Brown, J. H. Brewster, and H. Schechter, *J. Am. Chem. Soc.*, **76**, 467 (1954).

37. F. Price and L. P. Hammett, *J. Am. Chem. Soc.*, **63**, 2387 (1941).

38. G. Briegleb, *Naturwissenschaften*, **31**, 62 (1943).

39. M. Eigen and G. G. Hammes, *Adv. Enzymol.*, **25**, 1 (1963).

40. P. S. Tobias and F. J. Kézdy, *J. Am. Chem. Soc.*, **91**, 5171 (1969).

41. T. C. Bruice and B. Holmquist, *J. Am. Chem. Soc.*, **91**, 2993, 3003 (1969).

42. M. L. Bender, *J. Am. Chem. Soc.*, **73**, 1626 (1951).

43. M. L. Bender and R. J. Thomas, *J. Am. Chem. Soc.*, **83**, 4189 (1961).

44. A. J. Kirby, in *Comprehensive Chemical Kinetics*, Vol. 10, C. H. Bamford and C. F. H. Tipper, Eds., Elsevier, New York, 1972, Chap. II.

45. See footnote 11 in E. T. Kaiser, K.-W. Lo, K. Kudo, and W. Berg, *Biorg. Chem.*, **1**, 32 (1971).

# AUTHOR INDEX

Numbers in parentheses are reference numbers and show that an author's work is referred to although his name is not mentioned.

Numbers in *italics* indicate the pages on which the full references appear.

# SUBJECT INDEX

# CUMULATIVE INDEX, VOLUMES 1 to 4